T0230681

ANIMAL MODELS IN MEDICAL MYCOLOGY

Editor

Makoto Miyaji, M.D.

Professor
Department of Pathogenic Fungi Research Institute for Chemobiodynamics
Chiba University
Chiba, Japan

CRC Press
Taylor & Francis Group
Boca Raton London New York

CRC Press is an imprint of the
Taylor & Francis Group, an **informa** business

First published 1987 by CRC Press
Taylor & Francis Group
6000 Broken Sound Parkway NW, Suite 300
Boca Raton, FL 33487-2742

Reissued 2018 by CRC Press

© 1987 by CRC Press, Inc.
CRC Press is an imprint of Taylor & Francis Group, an Informa business

No claim to original U.S. Government works

This book contains information obtained from authentic and highly regarded sources. Reasonable efforts have been made to publish reliable data and information, but the author and publisher cannot assume responsibility for the validity of all materials or the consequences of their use. The authors and publishers have attempted to trace the copyright holders of all material reproduced in this publication and apologize to copyright holders if permission to publish in this form has not been obtained. If any copyright material has not been acknowledged please write and let us know so we may rectify in any future reprint.

Except as permitted under U.S. Copyright Law, no part of this book may be reprinted, reproduced, transmitted, or utilized in any form by any electronic, mechanical, or other means, now known or hereafter invented, including photocopying, microfilming, and recording, or in any information storage or retrieval system, without written permission from the publishers.

For permission to photocopy or use material electronically from this work, please access www.copyright.com (http://www.copyright.com/) or contact the Copyright Clearance Center, Inc. (CCC), 222 Rosewood Drive, Danvers, MA 01923, 978-750-8400. CCC is a not-for-profit organization that provides licenses and registration for a variety of users. For organizations that have been granted a photocopy license by the CCC, a separate system of payment has been arranged.

Trademark Notice: Product or corporate names may be trademarks or registered trademarks, and are used only for identification and explanation without intent to infringe.

Library of Congress Cataloging-in-Publication Data

Animal models in medical mycology.

Includes bibliographies and index.
1. Mycoses--Animal models. 2. Medical mycology--Research--Methodology. I. Miyaji, Makoto, 1937-
[DNLM: 1. Disease Models, Animal. 2. Mycoses. WC 450 A598]
RC117.A53   1987   616.9'69027   87-20869
ISBN 0-8493-5844-2

A Library of Congress record exists under LC control number: 87020869

Publisher's Note
The publisher has gone to great lengths to ensure the quality of this reprint but points out that some imperfections in the original copies may be apparent.

Disclaimer
The publisher has made every effort to trace copyright holders and welcomes correspondence from those they have been unable to contact.

ISBN 13: 978-1-315-89059-3 (hbk)
ISBN 13: 978-1-351-06969-4 (ebk)

Visit the Taylor & Francis Web site at http://www.taylorandfrancis.com and the
CRC Press Web site at http://www.crcpress.com

PREFACE

The prevalence of antibiotics, corticosteroids, and anticancer or immunosuppressive drugs and the progress of the medical treatments after World War II have saved the lives of many patients attacked by serious diseases and have contributed to the prolongation of the average life span of a human being. However, there still remain some problems to be solved. One of them is that of compromised hosts, which will be treated in this book. Opportunistic fungal infections are prone to occur in such patients and, recently, the number of patients with mycoses has been increasing more and more. Now, the study of the prevention of and therapy against mycoses has become an urgent problem.

In order to understand mycoses, fundamental studies, as well as clinical ones, are necessary. In particular, animal experiments are indispensable for the study of mycoses. Until now, various animal models have been developed in the field of medical mycology; however, there are few text books describing comprehensively the animal models of various fungal infections. Since graduating from the Chiba University School of Medicine in 1963, I have studied host defenses against fungal infections using animals, in particular mice. From my experience, I have keenly felt the necessity for this type of book. Fortunately, at this time we are able to publish a new book of *Animal Models in Medical Mycology* by courtesy of CRC Press, Inc. This book is published for postgraduate students and researchers studying host defense against fungal infections and those engaged in studying in vivo effects of antifungal agents. In addition, we believe it gives useful information to clinicians and medical technicians in hospitals dealing with patients with mycoses and fungi, respectively.

We would be glad if this book would contribute to the progress of the study of mycoses.

Makoto Miyaji, M.D.
December, 1985

THE EDITOR

Makoto Miyaji, M. D., is a Professor of the Department of Pathogenic Fungi, the Research Institute for Chemobiodynamics, Chiba University, Chiba, Japan.

Dr. Miyaji graduated in 1963 from the Faculty of Medicine, Chiba University, with a M. D. degree and received his Dr. Med. Sci. degree in 1986 from Chiba University.

Dr. Miyaji has served as an invited lecturer at the Faculty of Medicine, Chiba University.

Dr. Miyaji has published more than 100 scientific papers. His current major research interests include the defense mechanisms of host against fungal infections, ontogeny and phylogeny of pathogenic black yeasts, ecology of pathogenic fungi, and parasitic forms of pathogenic fungi.

CONTRIBUTORS

Elmer Brummer, Ph.D.
Research Associate
Santa Clara Valley Medical Center
Institute for Medical Research
Stanford University
San Jose, California

Karl V. Clemons, Ph.D.
Post Doctoral Scholar
Santa Clara Valley Medical Center
Institute for Medical Research
Stanford University
San Jose, California

Kazuko Nishimura, M.D.
Associate Professor
Department of Pathogenic Fungi
Research Institute for Chemobiodynamics
Chiba University
Chiba, Japan

Shohei Watanabe, M.D.
Professor
Department of Dermatology
Shiga University of Medical Science
Otsu, Japan

Hideyo Yamaguchi, M.D.
Professor
Research Center for Medical Mycology
Teikyo University
Tokyo, Japan

ACKNOWLEDGMENTS

Invaluable time and knowledge were generously contributed by the five authors of this book to whom I express my deepest gratitude. Also to CRC Press, Inc. for dedicated and skillful editorial and secretarial work.

Makoto Miyaji

TABLE OF CONTENTS

Chapter 1

EXPERIMENTAL FUNGAL INFECTIONS

M. Miyaji and K. Nishimura

TABLE OF CONTENTS

I. INTRODUCTION

Animal experiments are indispensable for the study of mycoses as well as other infectious diseases. Until now, various animal models of mycoses have been reported in the field of medical mycology (see Chapters 2 to 4). These animal models may be divided into three categories according to researcher's purposes. They are for the study of (1) the parasitic forms of fungi including the morphological transformation, (2) the defense mechanisms of host against fungal infections, and (3) the effect of antifungal agents.

In order to understand fungal infections, we should know well the following two aspects: how the parasitic forms of the causative agents and the cellular reactions against them change with the advancement of infection, and how the cellular reactions change according to the species of the causative agents.

Most microbiologists, including medical mycologists, seem to be convinced that fungal infections fall under a category of bacterial infection, and that the mechanisms invoked in fungal infections can be analyzed by the same measures adopted in analyzing those of the bacterial infections. However, keen-eyed medical mycologists have noticed that the fungal infections are delicately different from the bacterial ones. One characteristic in the fungal infections is the parasitic forms. Different from bacteria, some of the causative agents of mycoses undergo morphological transformation of their saprobic forms from mycelial to yeast-like (their parasitic form) in the host tissue. Namely, fungi adapt themselves to a severe circumstance such as the human body by transforming into their parasitic forms. This phenomenon is termed *dimorphism*[1] which is one of the important factors of the pathogenicity of some fungi.

The pathogenicity of dimorphic fungi is generally stronger than that of nondimorphic fungi, and therefore, the former are frequently the causative agents of the deep mycoses.[2-4] Even though some pathogenic fungi are able to change from a mycelial form to a spherical shape in the tissue, the fungi forming these spherical cells do not multiply by budding like those of *Histoplasma capsulatum, Sporothrix schenckii, Blastomyces dermatitidis,* and *Paracoccidioides brasiliensis,* nor by endospore formation like *Coccidioides immitis,* nor by fission as do the sclerotic cells of *Fonsecaea pedrosoi, Exophiala dermatitidis,* etc. When the circumstance becomes favorable for the spherical cells of these fungi, they germinate and extend their hyphae within the lesions. Dermatophytes[5] are an example of this type of fungi.

As mentioned above, to know how fungi change into their parasitic forms with the advancement of infection is an important matter in order to better understand fungal infections. To date, there have been many reports on the transformation of dimorphic fungi using tissue cultures[6-8] or various media[9,10] which are rich in nutrients and are incubated under special physical conditions such as a higher temperature, a higher partial pressure of CO_2, etc. All of these experiments were carried out in vitro. Usually, when we want to observe transformation in vivo, we inoculate intraperitoneally, subcutaneously, intracranially, or intravenously short hyphae, conidia, or hyphal fragments into animals, and follow the transformation with time by histological sections. However, it is difficult to observe in detail the transformation in these experiments because the inoculated hyphae are prone to scatter and disappear.

The authors[11] devised a method, "agar implantation", for the purpose of observing the parasitic forms of fungi in vivo, and with this method they have reported on the parasitic forms of several pathogenic fungi.[11-15] In brief, this method is as follows: an agar block embedded with young hyphae is implanted into the abdominal cavity of a mouse. It is removed after an adequate interval and observed directly with a light microscope. The advantages of the method are as follows: (1) hyphae are placed under a circumstance of the same temperature as that of a host; (2) most of the nutrients necessary for the growth of fungi are supplied by the peritoneal exudate of the host, as the implantation period is prolonged; (3) hyphae are protected from a direct attack of infiltrating cells such as polymorphonuclear leukocytes (PMNs), mononuclear cells (macrophages), etc. because they are embedded in the agar (since a mouse cannot digest the agar block, it is maintained in the abdominal cavity for a long period of time; therefore, it is possible to observe the transformation during an extended period of time); (4) hyphae are influenced only by the exudate containing enzymes of the infiltrating cells, antibody complement, and other humoral killing factors of the host; (5) we can observe the recovered agar block directly with a light microscope. Furthermore, it is possible to prepare histological sections or specimens for electron microscopy[13] from the agar block. By means of this method, the morphological transformation and the parasitic forms of pathogenic fungi can be observed and described in vivo.

II. AGAR IMPLANTATION METHOD

The agar implantation method is carried out as follows (Figure 1). Slide cultures are made by inoculating 1 cm² block (3 mm in thickness) of Sabouraud's conservation medium (polypeptone 1%, dextrose 0.5%, agar 1.5%) in its center with a pinhead-size bit of the mycelial growth of the study cultures. After covering the inoculated agar block with a cover glass, the preparation is incubated at 27°C for an adequate period. After the confirmation of good hyphal growth, a portion of the agar block, which contains young hyphal growth, is cut into 3 mm cubes with a sterile scalpel.

Mice of any strain and sex can be used in this protocol. Mice, weighing 18 to 21 g are prepared for agar implantation as depicted in Figure 1. A mouse is put into a beaker (500 to 1,000 m*l*) at the bottom of which absorbent cotton soaked with diethyl ether and covered with a sheet of aluminum foil has been placed. A magazine or jour-

1

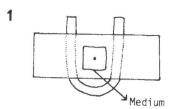

A fungus is inoculated at the center.

2   Medium   Cover glass

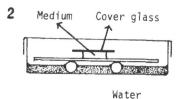

Water

3   Young hyphae

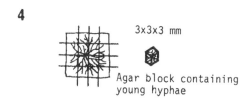

4

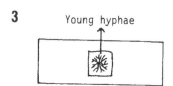

3x3x3 mm

Agar block containing
young hyphae

FIGURE 1.   Procedures of the agar-implantation method.

nal is used as a cover for the beaker. A second beaker (50 to 100 m*l*) containing absorbent cotton soaked with diethyl ether also is prepared. After the mouse in the large beaker loses consciousness, within 1 to 2 minutes, it is removed from the beaker and laid on its back on a tray covered with a sheet of aluminum foil. The small beaker is placed over the face as illustrated in Figure 1. The following operation must be done under sterile conditions. The abdominal wall is opened by an incision 0.5 to 1.0 cm in length. An agar block, prepared from the slide culture mentioned above, is inserted deeply into the abdominal cavity of the mouse and the incision is closed with sutures or clips. Usually the administration of antibiotics is not necessary to prevent infections. After surgery the mice are maintained under conventional conditions and are given common food and water *ad libitum*. The mice are killed at various intervals after implantation. Upon autopsy, a white gray nodule is easily found in the abdominal cavity of the mouse. The nodule is attached to the small intestine (Figure 2), peritoneum, liver, abdominal wall or fatty tissue around the urinary bladder or testes.

The granulomatous tissue covering the inoculum (Figure 3-1) is slit open with a scalpel blade and the exposed agar block (Figure 3-2) is put into a drop of a lactophenol solution on a glass slide. A cover glass is placed over it, pressed carefully and the specimen is observed directly using a light microscope. When a dangerous fungus, such

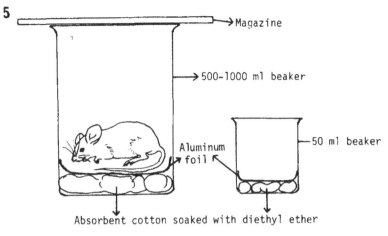

5
→ Magazine

→ 500-1000 ml beaker

→ 50 ml beaker

Aluminum
foil

Absorbent cotton soaked with diethyl ether

6

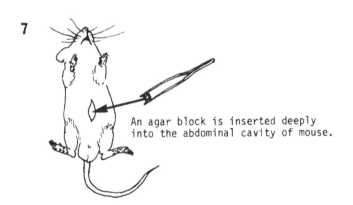

7

An agar block is inserted deeply
into the abdominal cavity of mouse.

FIGURE 1 (5-7)

as *Coccidioides immitis,* is used, the agar block removed from the abdominal cavity of mouse is fixed in a 10% formalin solution and, after fixation, is put into a drop of a formalin solution on a glass slide. When we want to prepare histological sections or specimens for electron microscopy, nodules are fixed in a 10% formalin solution or a 3% glutaraldehyde solution, respectively. Then, according to the conventional procedures, histological sections (Figure 4) or specimens for electron microscopy (Figure 5) are prepared.

III. CLASSIFICATION OF PATHOGENIC FUNGI ACCORDING TO THEIR PARASITIC FORMS

The pathogenic fungi are classified into two categories according to their parasitic forms. The first consists of the fungi which are mycelial in a saprobic life and spherical

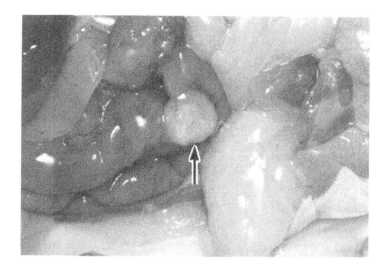

FIGURE 2.   Agar block is attached to the small intestine.

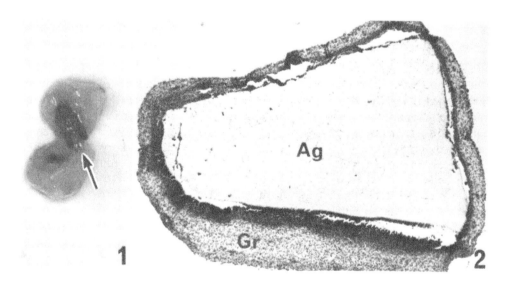

FIGURE 3.   Agar blocks recovered from the abdominal cavity of a mouse: (1) agar block (arrow) in a nodule and (2) histologic section of an agar block. Ag, agar block; Gr, Granulomatous tissue. (Periodic acid, Schiff.)

in a parasitic one (such as spherical cells, yeast-like cells, endospores, or sclerotic cells). The second consists of the fungi whose parasitic forms are similar to the saprobic forms.

A. Fungi Whose Parasitic Forms are Different from the Saprobic Ones
1. Fungi Changing from a Mycelial Form to Spherical or Yeast-Like Cells via the Stage of Arthroconidium
   Trichophyton rubrum, C. immitis and *Penicillium marneffei* apply to this case.

a. Trichophyton rubrum
   The transformation of *T. rubrum*, which is most frequently isolated from patients

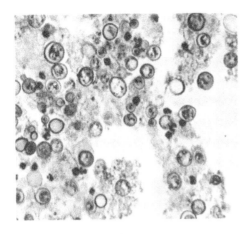

FIGURE 4. Yeast-like cells of *Blastomyces dermatitidis* in an agar block. (Periodic acid, Schiff.)

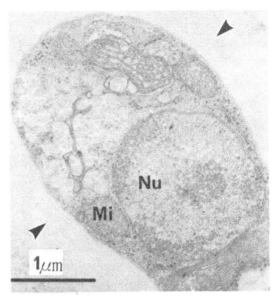

FIGURE 5. Electron micrograph of a yeast-like cell of *Blastomyces dermatitidis* in an agar block. The arrow heads indicate the cell wall. Nu, nucleus; Mi, mitochondrion.

with dermatophytoses, is explained. In general, it parasitizes the keratinized tissue such as the horny layer of the epidermis, hair, and nail, and rarely invades the dermis or subcutaneous tissue. It exists as a hyphal form with sparse septa (Figure 6-1) in the keratinized tissue in the early stages of infection. With the advancement of infection, the hyphae thicken and the number of septa increases (Figure 6-2). As a result, chains of arthroconidia (Figure 6-3) are formed. With marked inflammatory reactions, the chains break up into fragments, which swell and become spherical cells (Figure 6-4). The resistance of the spherical cells to infiltrating cells or killing factors in the exudate or serum is greater as compared with the resistance of the hyphal form.[5]

Spherical cells can be obtained in vitro by the following procedure. *T. rubrum* is inoculated into a Sakaguchi's flask containing 100 m*l* of brain-heart infusion broth (Difco) supplemented with 1% dextrose and incubated with reciprocal shaking at 37°C. Under these conditions, more than half of the cultures of *T. rubrum* change into chains of arthroconidia and spherical cells (Figure 7-1) within 3 to 4 weeks.[5] When these spherical cells are inoculated intraperitoneally into a mouse, they survive for more than 2 months in the necrotic center of the granulomatous nodule formed in the abdominal cavity of the mouse (Figure 7-2). However, when short hyphae are inoculated intraperitoneally, most are destroyed within 10 days after inoculation. Namely, the spherical cells are able to adapt to such an unusual circumstance like the abdominal cavity of a mouse. When the circumstances become favorable, the spherical cells may germinate (Figure 7-3) and extend their hyphae. These results seem to indicate that cultures of *T. rubrum*, which readily change from mycelial to spherical, have the potentiality to survive in the deep tissues. However, according to the recent data obtained when using the agar implantation method, some cultures of *T. rubrum* grow with a hyphal shape in the abdominal cavity of mouse for a long period of time.[16] Each of 13 isolates of *T. rubrum*, including 3 cultures isolated from patients with granuloma trichophyticum (in this disease *T. rubrum* grows in the subcutaneous tissue), was implanted in the abdominal cavity of a mouse. As a result, the cultures were divided into three groups according to their parasitic forms. Two cultures, isolated from tinea un-

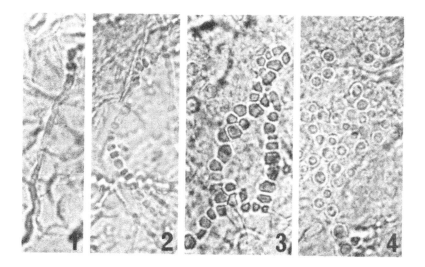

FIGURE 6.   Transformation of *Trichophyton rubrum* from hyphae to spherical cells (1 to 4). (KOH preparation.)

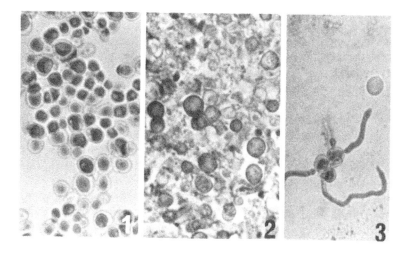

FIGURE 7.   Spherical cells of *Trichophyton rubrum:* (1) spherical cells produced in vitro (cotton blue); (2) spherical cells in a nodule formed in the abdominal cavity of a mouse (periodic acid, Schiff), and (3) germination of a spherical cell (cotton blue).

guium and tinea pedis, respectively, were destroyed in the abdominal cavity of a mouse within 5 weeks. Eight cultures, isolated from tinea cruris, tinea circinata and tinea pedis, transformed into chains of arthroconidia or spherical cells within 16 weeks. In the third group, the 3 cultures isolated from granuloma trichophyticum were viable in a hyphal form in the agar block for more than 26 weeks without changing to spherical cells.

Further studies are needed to clarify a dermatophyte's invasiveness of the deep tissue.

b. Coccidioides immitis

C. immitis is the causative agent of coccidioidomycosis which is the severest disease of all the mycotic infections.[17] It occurs after inhalation of the arthroconidia of *C. immitis*, which change into spherules packed with numerous endospores in the host tissue.

To date, there have been many reports on the transformation of *C. immitis* in vivo[18-20] and in vitro.[9,21]

When arthroconidia, prepared from slants, were inoculated intravenously into mice, the liver was severely affected by them, followed by the spleen, lung, and kidney in that order.[20] Most of the arthroconidia swelled and became spherules (7 to 33 μm in diameter) within 3 days after inoculation. A few spherules reached the midpoint stage of their parasitic cycle within 3 days. All the developmental stages, from spherical initials to released endospores, were found in the liver, spleen, and lung tissues on the 5th day. The inside of the spherule initials differentiated into two zones: An inner zone of a dense, curd-like substance and an outer zone of cytoplasm. The curd-like substance was polysaccharide in nature. As spherule development progressed, the curd-like substance disappeared, and the central zone became vacuolated. As the development progressed further, the cytoplasmic membrane and the inner layers of the cell wall began to invaginate at several points into the cytoplasm,[21] and divided it roughly into several large segments. The cleavage walls continued to branch towards the center of the spherule. As the cleavage walls repeatedly branched, they became thinner and divided the cytoplasm into progressively smaller segments. Then, the cytoplasm, made up of cleavage segments, separated at the relatively thick cleavage walls which had been formed at the initial stage of the cleavage process. As a result, the spherule was packed with relatively large packets of segmented cytoplasm. The surface of the packets was marked with shallow vertical and transverse furrows. When the packets were sectioned, they were observed to be composed of small segments. As soon as endospores developed in every cleavage segment, the sac walls and the cleavage walls within the sacs began to dissolve, and individual endospores were released within the spherule. As a result, the spherule was packed with a large number of free endospores. At the completion of the endosporulation process, the cell wall of the spherule ruptured, and the endospores were released (Figures 8 and 9).

When the parasitic cycle was observed by the agar implantation method (Figure 10),[15] arthroconidia implanted into the abdominal cavity of a mouse swelled and became immature spherules within 3 days, and after the 4th day, a few of the spherules had reached the stage of maturity. When hyphae were implanted, they swelled gradually with many septa appearing in them after the 2nd day forming chains of arthroconidia. Disjunctor cells were not formed.

c. Penicillium marneffei

According to Segretain,[22] hyphae of *P. marneffei* transformed to chains of arthroconidia in vivo, parted from each other and became yeast-like cells, which multiplied not by budding, but by fission.

2. Fungi Changing from a Mycelial Form to Spherical Cells, Sclerotic Cells, or Yeast-Like Cells via the Stage of Chlamydospore

The pathogenic dematiaceous fungi, *B. dermatitidis*, *P. brasiliensis* and *Microascus cirrosus* apply to this case.

a. Fonsecaea pedrosoi

As an example of the pathogenic dematiaceous fungi, the transformation of *F. ped-*

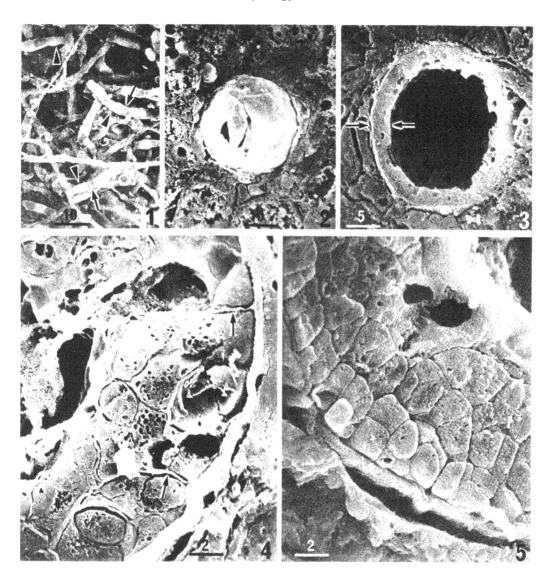

FIGURE 8.   Parasitic cycle of *Coccidioides immitis* in the liver of mice, scanning electron micrographs: (1) arthroconidia (arrow heads) and disjunctor cells (arrows); (2) young spherule in the liver; (3) cut section of a young spherule (arrows show the cytoplasm); and (4 and 5) cut sections of spherules (arrows show cleavage walls); (6) surface of sacs in a spherule; (7) cut section of sacs; (8) endospores in sacs, and (9) released endospores.

rosoi is described. *F. pedrosoi* is isolated most frequently from patients with chromomycosis. It exists in the cutaneous and subcutaneous tissues as sclerotic cells with thick cell walls, which are characteristic for the parasitic form of the pathogenic dematiaceous fungus.[2-4] These sclerotic cells are formed during the following process:[12] young hyphae of *F. pedrosoi,* implanted in the abdominal cavity of a mouse by the agar implantation method, swelled irregularly. Terminal and intercalary chlamydospores were formed after the 14th day of the implantation. As the implantation period progressed, the number of chlamydospores increased, followed by a rounding up, and became chains of spherical cells. These chains of spherical cells parted from each other and became sclerotic cells within 20 to 30 days after implantation (Figure 11). Sclerotic cells multiply in two ways. The first is by germination and extension of hyphae under

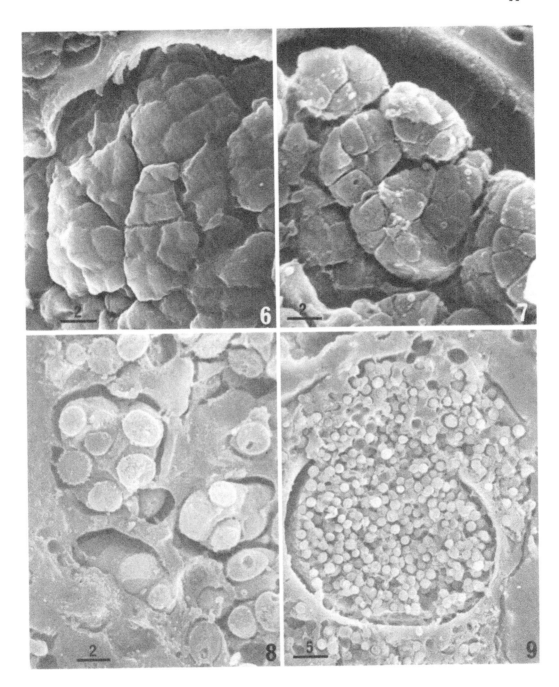

FIGURE 8 (6—9)

favorable environmental conditions. The second is that sclerotic cells with septa (Figure 12) are split into cells, which become new sclerotic cells, respectively.

b. *Blastomyces dermatitidis*

Young hyphae of *B. dermatitidis* implanted into the abdominal cavity of a mouse formed intercalary and terminal chlamydospores within 9 days (Figure 13-1).[13] These chlamydospores were observed at the periphery of the agar block. Fourteen days after

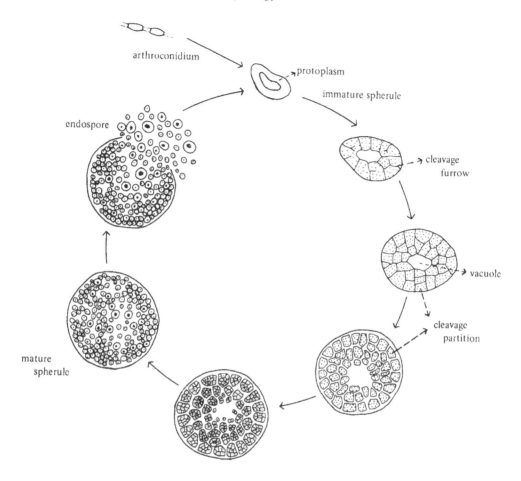

FIGURE 9.   Parasitic cycle of *Coccidioides immitis*.

implantation the chlamydospores swelled and became spherical cells which reproduced daughter cells by single budding (Figure 13-2,3). The daughter cells were attached to the parent cells by a wide base (Figure 14). Occasionally, yeast-like cells occurred by direct budding from hyphae (Figure 13-4).

According to Howard and Herndon,[7] mycelial fragments of *B. dermatitidis* change to a "chain of oidia" which part from each other and become yeast cells in tissue culture. Clinically, it is believed that blastomycosis occurs after inhalation of conidia of this fungus, and the primary lesion is produced in the lung.[2-4] However, it is unclear whether or not the inhaled conidia reproduce their daughter cells by direct budding. According to the observations of Garrison and Boyd[23] in vitro, yeast-like cells (bud-like structures) were budded directly from a germtube formed by germination of a conidium at 37°C.

c. *Paracoccidioides brasiliensis*

In *P. brasiliensis*, transformation is the same as that of *B. dermatitidis*. Within 7 days after implantation young hyphae transformed to chains of chlamydospores and became spherical cells (Figure 15-1), which after the 14th day produced daughter cells by multiple budding (Figure 15-2); Salazar and Restrepo reported the same conversion process in vitro.[24] However, different from *B. dermatitidis*, the daughter cells were attached to the parent cells by a narrow attachment. Clinically, it is also believed that

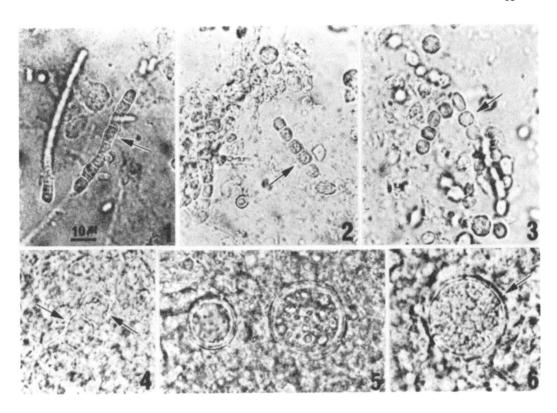

FIGURE 10.   Parasitic cycle of *Coccidioides immitis* observed by the agar implantation method: (1 to 4) arrows indicate arthroconidia; (5) two immature spherules, and (6) arrow indicates a spherule (KOH preparation.) (From Miyaji, M. and Nishimura, K., *Mycopathologia*, 90, 122, 1985. With permission.)

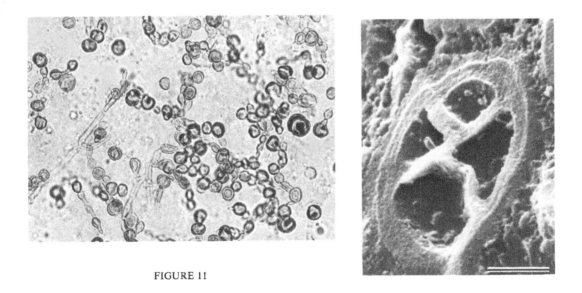

FIGURE 11

FIGURE 12

FIGURE 11.   Sclerotic cells of *Fonsecaea pedrosoi* formed in the abdominal cavity of a mouse. (KOH preparation.)

FIGURE 12.   Sclerotic cell of *Exophiala dermatitidis* divided into four parts by the septa. (Scanning electron micrograph. Bar indicates 10 μm.)

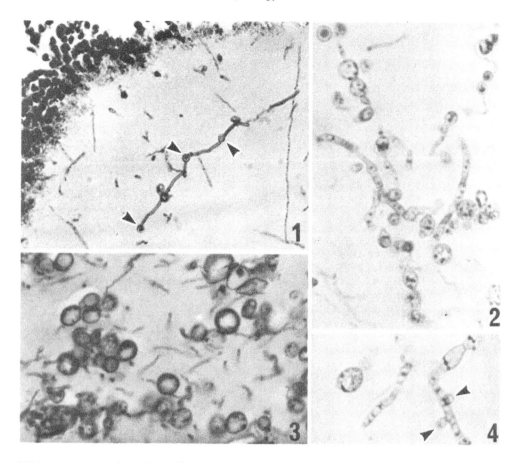

FIGURE 13.   Transformation of *Blastomyces dermatitidis* observed by the agar implantation method: (1) Chlamydospore formation. The arrow heads indicate chlamydospores, (2) chlamydospores and spherical cells, (3) clumps of spherical cells, and (4) spherical cells produced from a hypha by direct budding. (Periodic acid, Schiff.)

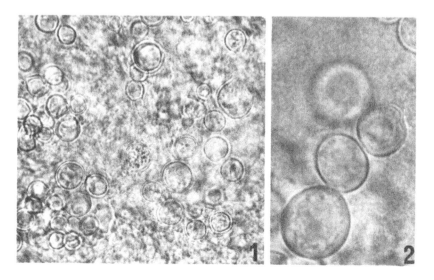

FIGURE 14.   Spherical cells of *Blastomyces dermatitidis* observed directly with a light microscope: (1) cells reproduce daughter cells by single budding; (2) daughter cells are attached to the parent cells by a wide base. (KOH preparation.)

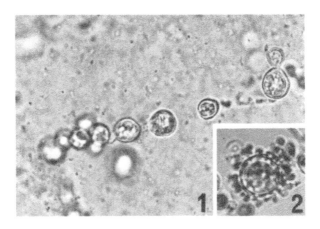

FIGURE 15.   Transformation of *Paracoccidioides brasiliensis*
observed by the agar implantation method: (1) chlamydospore
formation and (2) spherical cell with multiple budding. (KOH
preparation.)

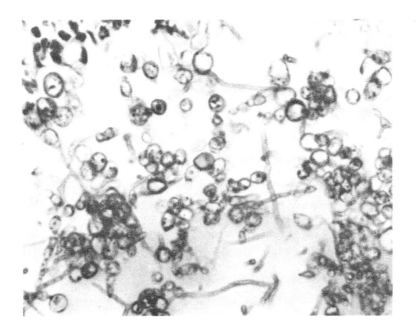

FIGURE 16.   Spherical cells of *Microascus cirrosus.* (Periodic acid, Schiff.)

paracoccidioidomycosis occurs after inhalation of conidia.[4] However, it is unclear
whether or not the inhaled conidia reproduce their daughter cells by direct budding.

d. *Microascus cirrosus*

Young hyphae of *M. cirrosus*[14] implanted into the abdominal cavity of a mouse
swelled gradually and on the 4th week intercalary and terminal chlamydospores were
produced, forming chains of spherical cells, which reproduced daughter cells by multiple budding, forming clumps of spherical cells (Figure 16). This transformation is the
same as that of *P. brasiliensis*. The young hyphae of *M. cinereus* and *M. trigonospo-
rus*[14] changed to chains of spherical cells like those of *M. cirrosus*. However, they did
not multiply by budding.

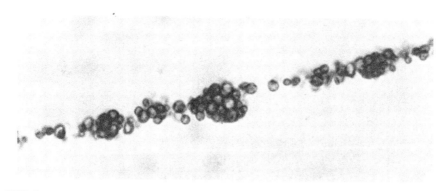

FIGURE 17.   Clumps of yeast-like cells of *Sporothrix schenckii* around a hypha which disappeared with the advancement of implantation period. (Periodic acid, Schiff.)

3. Fungi Changing from a Mycelial Form to a Yeast Form by Direct Budding from Hyphae

In this case yeast cells are produced chiefly by direct budding from hyphae. *Sporothrix schenckii, Histoplasma capsulatum* variant *capsulatum,* and *H. capsulatum* var. *duboisii* apply to this case.

a. Sporothrix schenckii

When young hyphae of *S. schenckii* were implanted into the abdominal cavity of a mouse, yeast cells were consecutively produced from hyphae.[11] Subsequently, the yeast cells produced daughter cells by multiple budding. Eventually, as shown in Figure 17, clumps of yeast cells were formed around the hypha, which had disappeared with the advancement of the implantation period.

According to Howard,[8] a mycelial form of *S. schenckii* transforms to a yeast form in two ways. The first way is that club-shaped cells are produced on the hyphal wall and at the tips of lateral branches when the mycelial form of the fungus is inoculated into a tissue culture of mouse mononuclear cells. The fungal cells bud repeatedly, forming clumps of yeast cells. The second way is that condensation of hyphal cytoplasm occurs, followed by the formation of chains of oval or elongated cells, termed "oidia", which part from each other and produce yeast cells. The yeast cells produce daughter cells by multiple budding. It is another interesting problem whether conidia produce yeast cells by direct budding or not. According to Maeda,[25] the plasma membrane of the conidia is definitely different from that of the yeast cells at an ultrastructural level, and the conidia never produce daughter cells by direct budding but first germinate and extend germtubes, and then yeast cells occur from the germtubes by direct budding.

Clinically, it is difficult to find fungal elements of *S. schenckii* in histological sections prepared from biopsy materials.[2-4] However, there are various shapes of fungal elements in the scales and crusts in sporotrichotic lesions.[26]

b. Histoplasma capsulatum var. capsulatum

Young hyphae of *H. capsulatum* var. *capsulatum* implanted into the abdominal cavity of a mouse swelled gradually. A condensation of the cytoplasm occurred in some parts of the hyphae, and at the same time intercalary and terminal chlamydospores were formed 2 days after implantation. Three days after implantation, the width of hyphae and the diameter of the chlamydospores were 2 to 2.5 and 3 to 5 μm, respectively. Ten days after implantation moniliform hyphae were formed in some parts of the hyphae. After the 14th day, more than half of the hyphae and chlamydospores had

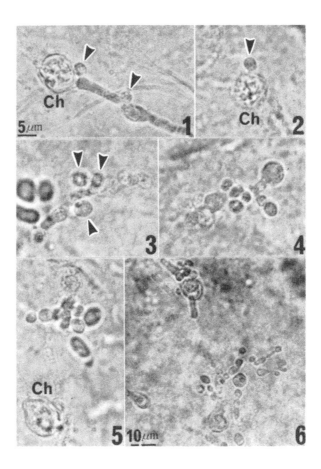

FIGURE 18. Transformation of *Histoplasma capsulatum* observed by the agar implantation method. Arrow heads indicate yeast-like cells; Ch, chlamydospore. (1) Yeast-like cells produced from a hypha by direct budding. (2) Yeast-like cells produced from a chlamydospore by direct budding. (3) Yeast-like cells produced from a thickened hypha by direct budding. (4 and 5) Small clumps of yeast-like cells. (6) Yeast-like cells formed from moniliform hyphae. (KOH preparation.)

lost their cytoplasm, and within 30 days, most of them were broken down and disappeared gradually. Only a few short hyphae, chlamydospores, and moniliform hyphae survived in the agar block, and produced yeast cells by direct budding.[11] These yeast cells produced daughter cells by multiple budding (Figure 18).

To date, there have been many reports dealing with the transformation of *H. capsulatum* var. *capsulatum*. In short, there are 3 mechanisms involved in their transformation: (1) yeast cells are produced from hyphae by direct budding; (2) hyphae change to moniliform hyphae, from which yeast cells are produced by direct budding; (3) yeast cells are produced from "stalked yeast cells" on hyphae. Pine and Webster[10] demonstrated the first transformation by using Pine's medium; in the observation of the agar implantation method, this is the main mechanism. The second transformation was indicated by Pine and Webster,[10] Moore,[27] and Howard[6] by using Pine's medium, a chorio-alantoic membrane, and a tissue culture of mouse mononuclear cells, respectively. The third was described by Pine and Webster.[10] According to them, "stalked yeast cells" develop in a manner similar to that of microconidium production. The stalked yeast cells produce yeast cells by direct budding.

FIGURE 19.   Parasitic form of *Aspergillus fumigatus* in the lung. (Periodic acid, Schiff.)

Next, the transformation from microconidia to yeast cells is discussed. Pine and Webster,[10] Dowding,[28] and Goos[29] reported that yeast cells are produced by direct budding from microconidia. According to Garrison's study using transmission electron microscopy,[30] microconidia germinate and become "yeast mother cells" which bud yeast-like daughter cells.

Three mechanisms were reported for the transformation from macroconidia to yeast cells, one is by occurrence of endospore-like cells within the macroconidia, and another is by swelling and separation of tubercules from macroconidia.[28] The former was observed by Procknow et al.,[31] Moore,[27] and Bradt.[32] According to them, when the macroconidia, containing endospore-like cells, mature, their cell walls rupture, and a great number of endospore-like cells are released from them. On the other hand, Garrison[30] and Pine and Webster[10] demonstrated a possibility of the transformation from macroconidia to yeast cells. They supported the process that macroconidia extend germtubes which bud yeast-like cells directly, or which change to moniliform hyphae which produce yeast cells by direct budding.

According to Pine and Webster[10] and Howard,[6] a mycelial form of *H. capsulatum* var. *capsulatum* transformed to yeast cells within 2 days in vitro. In the agar implantation method, hyphae transformed to yeast cells after the 10th day post implantation.

The mycelial form of *H. capsulatum* var. *duboisii* transformed to a yeast form mainly by direct budding from hyphae as well as in *H. capsulatum* var. *capsulatum*, when young hyphae of this fungus were implanted into the abdominal cavity of a mouse.[113]

B. Fungi Whose Parasitic Forms Are Similar to the Saprobic Forms

Parasitic form is mycelial as is the saprobic form — Most of the etiolotic agents of opportunistic fungal infections fall into this category. For example, the parasitic form of *Aspergillus fumigatus* is usually a hyphal form with characteristic dichotomous branching in the tissue (Figure 19). *Absidia, Mucor,* and *Rhizopus* exist as nonseptate hyphae in vivo as well as in vitro. *Fusarium* and *Pseudallescheria boydii* are observed as septate hyphae both in vivo and in vitro.

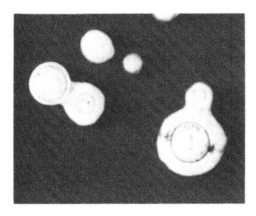

FIGURE 20.   *Cryptococcus neoformans* coated with polysaccharide in India ink preparation.

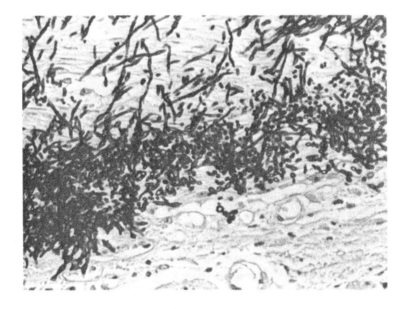

FIGURE 21.   Parasitic form of *Candida albicans* in the submucosal layer of the esophagus. It consists of both pseudohyphae and blastoconidia. (Periodic acid, Schiff.)

Parasitic form is a yeast form as is the saprobic form — *Cryptococcus neoformans* is an example of this type. It is coated with an acid-polysaccharide capsule in vivo and in vitro (Figure 20).

Parasitic form consists of both mycelial and yeast forms as is the saprobic form — *Candida* spp. are examples of this type. Yeast cells of *Candida* spp. are produced from pseudohyphae, true hyphae, or blastocomidia by direct budding (Figure 21).

C. Conclusions

The transformation of pathogenic fungi observed by the agar implantation method is summarized in Figure 22.

A. Fungi whose parasitic forms are different from the saprobic forms

    1. Fungi changing from a mycelial form to spherical or yeast-like cells *via* the stage of arthroconidium

    2. Fungi changing from a mycelial form to spherical cells, sclerotic cells or yeast-like cells *via* the stage of chlamydospore

    3. Fungi changing from a mycelial form to a yeast form by direct budding from hyphae

B. Fungi whose parasitic forms are similar to the saprobic forms

    1. Parasitic form is mycelial as is the saprobic form

    2. Parasitic form is a yeast form as is the saprobic form

    3. Parasitic form consists of both mycelial and yeast forms as is the saprobic form

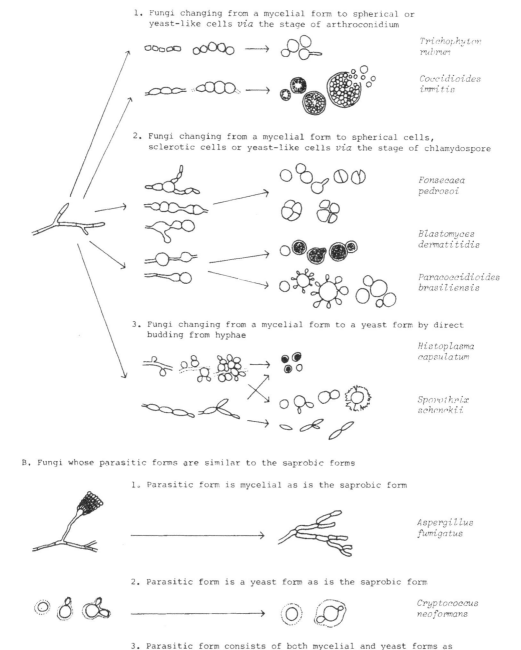

Trichophyton rubrum

Coccidioides immitis

Fonsecaea pedrosoi

Blastomyces dermatitidis

Paracoccidioides brasiliensis

Histoplasma capsulatum

Sporothrix schenckii

Aspergillus fumigatus

Cryptococcus neoformans

Candida albicans

FIGURE 22.   Classification of the pathogenic fungi according to their transformation and parasitic forms.

IV. TISSUE RESPONSES AGAINST FUNGAL INFECTION

In order to understand the various fungal infections, we have to know the histopathologic course of each fungal infection. As is well-known, the tissue responses change variously according to the species of fungi and the stage of infection. To date, various animal models have been used for studies of the mycoses. Among them the mouse is most frequently used because there are many hereditarily pure mouse strains which are useful for the analysis of obtained experimental data. In addition, mice are easy to deal with.

A chief histopathological characteristic of the deep mycoses is a granulomatous inflammatory reaction which plays an important defensive role against fungal infections.[33] Therefore, it is necessary for the understanding of the fungal infections to know the time course of granuloma formation and the killing functions of the granuloma. There are two questions with regard to the granuloma formation and the killing functions, one is whether or not cell-mediated immunity relates to the granuloma formation against fungal infections, and the other is whether or not the granuloma has the capability to kill the fungi within the granuloma without any exertion of cell-mediated immunity. Congenitally athymic nude (nu/nu) mice are suitable for the study of this relationship because of the defective functions of their T-lymphocytes.[34-37] To date, nu/nu mice have been used for the analysis of host defense mechanisms against various fungal infections; however, most of the reports focused on the "functions" cell-mediated immunity played in host defense against fungal infections and not on their histopathology.

In this chapter, the change of histopathological features with the advancement of infection are described based on the experimental data obtained from nu/nu mice,* their heterozygous littermates (nu/+mice) from BALB/c background, and other immunocompetent mice infected with various fungi.

A. Fungal Infections in Which Mononuclear Cells Play a Leading Role in Host Defense

1. Cryptococcus neoformans Infection

C. neoformans is the causative agent of cryptococcosis, an opportunistic fungal infection. Because the fungus has a predilection for the central nervous system,[2-4] more than half of the patients with cryptococcosis have been diagnosed as having cryptococcal encephalitis or cerebrospinal meningitis.

To date, there have been few reports[38-40] on murine cryptococcosis using nu/nu mice. These indicate that nu/nu mice are more susceptible to *C. neoformans* than are nu/+ mice. However, some explanation is needed with regard to these results. According to Cauley and Murphy,[38] the susceptibility of nu/nu mice to *C. neoformans* was less than that of nu/+ mice until the 14th day after inoculation. Thereafter, the nu/nu mice were affected more severely than were the nu/+ mice. They postulated that this early resistance was due to the increased function of macrophage in nu/nu mice.

Virulence of *C. neoformans* to the mouse is relatively strong. According to our data,[113] most ddY mice when inoculated intravenously with 5×10^6 yeast cells died within 1 week. Usually in the infected mice, the liver and brain were affected severely. The histopathological features in the brain of the mouse are usually cystic (Figure 23) and granulomatous in the liver[40,41] (Figure 24-1). These two types of lesions are defi-

---

* Because of the defective functions of T-lymphocytes, nu/nu mice have to be dealt with carefully. However, it is not necessary to deal with them under completely sterilized conditions. A glove box, sterilized with formalin vapour before its use, is placed in a conventional room isolated from other experimental animals. Nu/nu mice are maintained in it, while being given sterilized cages, bed, food, and water.

FIGURE 23.   Cystic lesion in the brain of a nu/+ mouse inoculated intravenously with *Cryptococcus neoformans*. (Periodic acid, Schiff.)

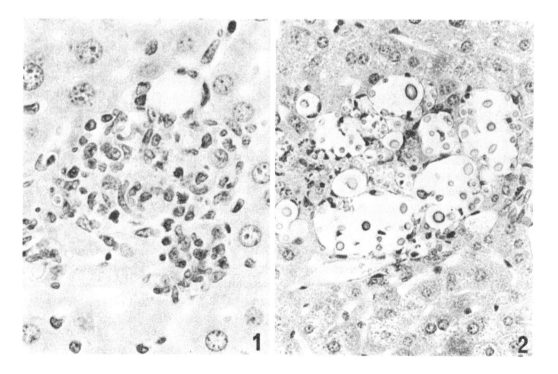

FIGURE 24.   Histopathological characteristics of the livers of mice inoculated intravenously with *Cryptococcus neoformans*: (1) granulomatous lesion, nu/+ mouse, and (2) cystic lesion, nu/nu mouse. (Periodic acid, Schiff.)

Table 1

NUMBER OF MICROFOCI IN 35 mm² OF LIVER SECTIONS

	Days after inoculation									
	2	3	4	6	8	10	14	18	21	25
Nu/+mice	0	0	3	6	4	19	21	9	5	6
Nu/nu mice	0	0	7	6	10	22	42	↑	↑	↑

Note: Each mouse was inoculated intravenously with 10^5 yeast cells of *C. neoformans* RIB-12. ↑: numerous.

nitely different from each other. The cystic lesion appears without any cellular responses, while the granulomatous one is formed as a result of various cell infiltrations. Thus, it is very interesting that these two types of lesions are simultaneously observed in a patient or an animal infected with *C. neoformans*.

In nu/+ or other immunocompetent mice inoculated intravenously with *C. neoformans*,[40] PMNs infiltrates were induced by the cryptococci during the early stages of infection in the liver. Seven days after inoculation, mononuclear cells (macrophages) accumulated at these foci changing the histopathological features from pyogenic to granulomatous. After the 14th day, cryptococci, which stained poorly by the periodic acid-Schiff (PAS) procedures, were gradually destroyed in the granulomatous lesions. On the other hand, the histopathological findings in the liver of nu/nu mice,[40] infected intravenously with *C. neoformans,* were cystic lesions with a large number of cryptococci (Figure 24-2). No cellular responses were induced in them and the cryptococci continued to multiply in the cystic lesions. The number of granulomatous lesions in the liver section of immunocompetent mice reached a peak on the 14th day and decreased thereafter, while in the nu/nu mice the number of the cystic lesions continued to increase with the advancement of infection (Table 1). The number of colony-forming units (CFU) in the livers of the nu/+ mice reached a peak on the 12th day and gradually decreased thereafter (Figure 25), whereas the CFU in the livers of nu/nu mice continued to increase.[42]

The histopathological features in the brains of both nu/nu and nu/+ mice were cystic, with the number of lesions continuing to increase with the advancement of infection (Table 2). Namely in the brain, regardless of the immunocompetence of the mouse, it cannot mount any defensive measures against *C. neoformans* infection. Table 3 demonstrates a relationship between the granulomatous and cystic lesions in the liver of both nu/+ and nu/nu mice. The granulomatous lesions are formed in the livers of the nu/+ mice inoculated intravenously with *C. neoformans* and the cystic ones appear in the livers of the nu/nu mice. When nu/nu mice are adoptively transferred with lymph node cells of nu/+ mice prior to inoculation of the fungus, granulomatous lesions are formed. When carrageenan treatment is used as a blockade of macrophages, cystic lesions appear instead of granulomatous lesions in the livers of the nu/+ mice.[40] These results appear to indicate that at least two steps are required for granuloma formation against *C. neoformans* infection. One is the phagocytosis by sessile macrophages and the second is the exertion of T-cell functions. Therefore, it may be postulated that granulomatous lesions are readily formed in the organs rich in the sessile macrophages, and cystic lesions in those poor in them. In immunocompetent mice inoculated intravenously with *C. neoformans* and killed 10 days after inoculation, granulomatous lesions were readily formed in the liver, spleen, lymph nodes, lung, thymus, etc., while the cystic type appeared in the brain, heart, pancreas, kidney, etc.[41]

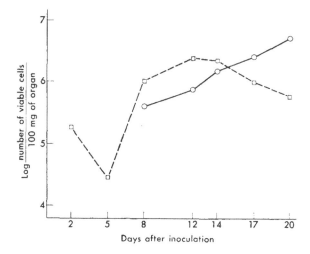

FIGURE 25.   Colony-forming units in the brain and liver of mice inoculated intravenously with *Cryptococcus neoformans.* ○, brain; □, liver. (From Watabe, T., Miyaji, M., and Nishimura, K., *Mycopathogia,* 86, 115, 1984. With permission.)

Table 2

AREA AND NUMBER OF CYSTIC LESIONS IN 80 mm² OF BRAIN SECTIONS

	Days after inoculation							
	2—4	6	8	10	14	18	21	25
Nu/+mice								
Area of cystic lesion (mm²)	0	0.01	0.06	0.13	0.44	2.94	3.17	4.05
Number of cystic lesion	0	1	2	3	5	13	11	15
Nu/nu mice								
Area of cystic lesion (mm²)	0	0.02	0.33	0.21	1.10	4.06	3.92	4.71
Number of cystic lesion	0	3	9	5	10	28	20	23

Note: Each mouse was inoculated intravenously with 10^5 yeast cells of *C. neoformans* RIB-12.

Table 3

CHIEF HISTOPATHOLOGIC FINDINGS OF THE LIVER OF MICE AFTER INTRAVENOUS INOCULATION OF *C. NEOFORMANS*

Mice	Granulomatous lesion	Cystic lesion
Nu/+	+	−
Nu/nu	−	+
Nu/nu mice transferred with lymph node cells of nu/+ mice	+	−
Nu/+ mice injected with 0.5 mg of carrageenan	−	+

Note: Mice were sacrificed 10 days after inoculation.

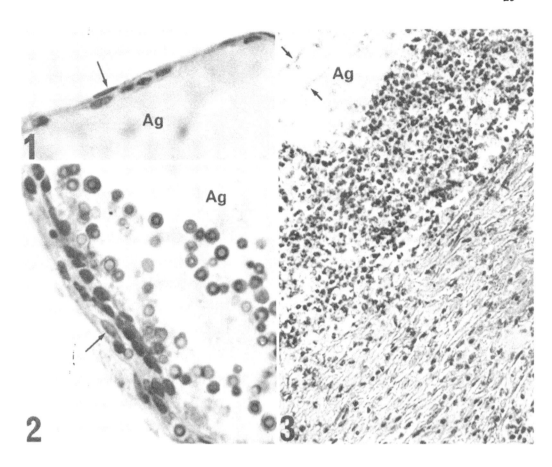

FIGURE 26.   Cellular responses to *Cryptococcus neoformans* embedded in agar blocks: (1 and 2) mild cellular responses to the agar blocks containing *C. neoformans* with a polysaccharide capsule, arrows indicate the layer of peritoneal macrophages, Ag, Agar block; (3) marked cellular responses to the agar block containing the teleomorph of *C. neoformans* (arrows). The agar block is surrounded by a thick inflammatory tissue consisting of two zones: The inner zone consists of PMNs and their debris, and the outer one consists of nonspecific granulomatous tissue. (Periodic acid, Schiff.) (From Miyaji, M. and Nishimura, K., *Mycopathologia*, 76, 152, 1981. With permission.)

The polysaccharide capsule that lies around cryptococci also plays an important role in *C. neoformans* infection. It may have an effect of inhibiting the migration of PMNs towards cryptococci.[41] When an agar block containing cryptococci coated with a thick polysaccharide capsule was implanted into the abdominal cavity of immunocompetent mice by the agar implantation method, a mild cellular reaction occurred in the agar block (Figure 26-1 and 2), while a remarkable purulent reaction was induced in another agar block containing the mycelial form of *C. neoformans* (teleomorph) which was formed by sexual mating (Figure 26-3). As is well-known, the mycelial form is not coated with the polysaccharide material.

When an immunocompetent mouse was inoculated intravenously with both *Aspergillus fumigatus* and *C. neoformans* (dual infection), two types of lesions (such as cystic and purulent lesions) appeared on the brain tissue at the same time (Figure 27).[41] The cystic lesion was formed by *C. neoformans* and the purulent reaction occurred against *A. fumigatus* infection. Even though the former lesion is very close to the latter as shown in Figure 27, PMNs are never induced by *C. neoformans*. This result seems to indicate that a mouse can distinguish other pathogens which are invading the brain tissue already infected with *C. neoformans*, and the PMNs migrate towards these pathogens. However, the mouse cannot detect the cryptococci as an invader.

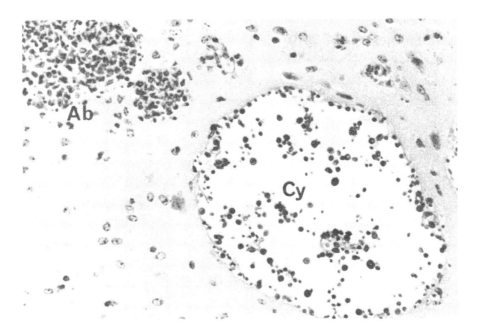

FIGURE 27.   Dual infection with *Cryptococcus neoformans* and *Aspergillus fumigatus* in the brain of a mouse. A cystic lesion (Cy) is formed by *C. neoformans* and abscesses (Ab) are formed by *A. fumigatus*. (Periodic acid, Schiff.)

As mentioned above, the granulomatous lesion seems to play a leading defensive role against *C. neoformans* infection. Therefore, it is an important problem whether the cystic lesion formed in the mouse brain is capable of changing into a granulomatous type. According to Grosse et al.,[43] a slow and weak glia-reaction occurred in some of the cystic lesions by the beginning of the 3rd week. When the authors[42] observed the brains of mice infected with *C. neoformans* for 55 days, cellular response did not occur in the cystic lesions formed in the brain parenchyma, even though cellular responses occurred in the leptomeninges, subarachnoid spaces, and chorioid plexuses. Thus, in the brain parenchyma, the mouse has no defensive measures against *C. neoformans* infection.

2. Sporothrix schenckii Infection

S. schenckii, which is the causative agent of sporotrichosis, characteristically affects the skin, subcutaneous tissue, and lymphatics. The clinical appearances are nodular lesions and indolent ulcers.[2-4] The histopathological characteristics of the lesions are granulomatous inflammatory reactions. There have been four published reports on experimental murine sporotrichosis using nu/nu mice.[44-47] According to Shiraishi et al.,[44] the authors,[46] and Dickerson et al.,[47] the susceptibility of nu/nu mice to *S. schenckii* was higher than that of nu/+ mice and they concluded that cell-mediated immunity played a leading role in host defense against *S. schenckii* infection. On the other hand, Hachisuka[45] insisted that the defense mechanisms of mice consisted of both the phagocytosis by PMNs and the defense led by cell-mediated immunity.

In experimental sporotrichosis, a histopathological feature during the early stages of infection is a purulent inflammatory reaction, while during the latter stages, a granulomatous inflammatory response appears. These histopathological findings seem to indicate that the granuloma might play an important defensive role against *S. schenckii* infection. Immunologically, a delayed-type skin test (sporotrichin reaction) for pa-

Table 4

NUMBER OF LESIONS AND YEAST CELLS DETECTED IN 35 MM² OF LIVER SECTIONS OF MICE INOCULATED WITH *S. SCHENCKII*

	Days after inoculation					
	3	7	10	14	20	30
nu/ + mice	32[a]	172	248	39	13	4
	(169)[b]	(180)	(144)	(7)	(0)	(0)
nu/nu mice	24	148	268	399	>500	>500
	(128)	(212)	(318)	(612)	(numerous)	(numerous)

Note: Each mouse was inoculated intravenously with 10⁶ yeast cells of *S. schenckii* Sp-1.

[a]   Number of lesions.
[b]   Number of yeast cells in the lesions.

tients with sporotrichosis shows a positive reaction almost without exception,[48] while an appraisal by serological diagnosis is low compared with the delayed-type skin test. According to Shiraishi et al.,[44] the CFU in the liver, spleen, and kidneys of nu/nu mice, inoculated intravenously with *S. schenckii,* continued to increase even after the 14th day of infection, and a sporotrichin reaction in nu/+ mice exhibited a positive 14 days after inoculation.

When nu/nu, nu/+, or ddY mice were inoculated intravenously with 10⁶ yeast cells of *S. schenckii,* the liver was affected most severely, regardless of the strain of mouse, followed by the subcutaneous tissue of the tail (at the site of the injection), spleen, and kidneys, in this order.[46] Histopathologically in the liver of the nu/+ mouse, the yeast cells, engulfed by Kupffer's cells, were observed sporadically 2 days after inoculation. On the 3rd day, PMNs were elicited by the yeast cells and formed microabscesses. On this day, mononuclear cells began to accumulate at the foci, and on the 7th day, most of the lesions had changed to a granulomatous type, in which the yeast cells were positively stained by PAS. On the 10th day, the number of granulomatous lesions reached a peak (Table 4), and thereafter, the yeast cells began to be destroyed within the granulomatous lesions. On the 14th day, the number of the granulomatous lesions decreased abruptly with some of them becoming fibrous. The course of infection in the nu/nu mice was similar to that in the nu/+ mice until the 10th day. In the livers of the nu/nu mice microabscesses were formed during the early stages of infection similar to those observed in the nu/+ mice. On the 7th day, less than half of the lesions changed to granulomatous and on the 10th day, most of the lesions became granulomatous. About a 2 day delay was observed in granuloma formation in the nu/nu mice in comparison with that in the nu/+ mice. Yeast cells still retained a good shape in the granulomatous lesions in nu/nu mice. However, after the 10th day there was a definite difference in the killing function between the granulomatous lesions of the nu/nu and the immunocompetent mice. In the livers of the nu/nu mice the number of granulomatous lesions continued to increase with the advancement of infection (Figure 28). The yeast cells were not killed in the granulomatous lesions and continued to multiply. Namely, different from the granulomatous lesions of the nu/+ mice or other immunocompetent mice, those formed in the nu/nu mice did not have the capacity to kill the yeast cells in them,[46] thus indicating that a cooperation with T-cell functions was necessary for the killing of the yeast cells. Immunologically, a significant macrophage migration inhibition factor (MIF) response developed 11 days after inoculation in the nu/+ mice,[46] and this day nearly coincided with the day when the yeast cells in the

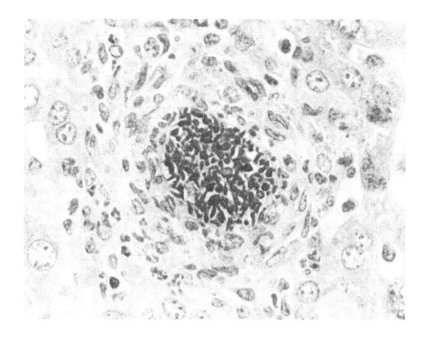

FIGURE 28.   Granulomatous lesion formed in the liver of a nu/nu mouse inoculated intravenously with *Sporothrix schenckii*. A clump of yeast cells is observed at the center. (Periodic acid, Schiff.)

granulomatous lesions began to be destroyed. On the other hand, it seems that humoral immunity did not play an important role in the defense mechanisms of mice because an agglutinating antibody was not detected for 20 days.

In summary, T-cell functions are indispensable for the killing of yeast cells of *S. schenckii* in granulomatous lesions.

3. Histoplasma capsulatum var. capsulatum Infection

H. capsulatum var. *capsulatum* is the causative agent of histoplasmosis capsulati which is a common and important systemic fungal disease in men and animals.[4] In disseminated histoplasmosis capsulati, the fungus chiefly affects parts of the reticulo-endothelial system, such as the liver, spleen, lymph nodes, etc., which results in marked proliferation of mononuclear cells. Even though PMNs are observed in the granulomatous lesions, they do not play an important role in the defense mechanisms of the hosts.[49] Many studies of host defense against *Histoplasma capsulatum* have been reported. According to these,[50-54] cell-mediated immunity plays a crucial role in the host's defense mechanisms against the fungal infection. Therefore, nu/nu mice are quite suitable for the studies of experimental histoplasmosis capsulati. Williams et al.[55] reported that nu/nu mice were affected more severely by *H. capsulatum* in comparison with nu/+ mice. They emphasized the importance of cell-mediated immunity in the host's defense against the fungal infection, too. They observed large numbers of yeast cells with no cellular infiltrate in the liver tissue of the nu/nu mice 50 days after inoculation. Especially, cell-mediated immunity played a leading role in eliminating the pathogen during the first 2 weeks of the infection in nu/+ mice.[49]

When nu/nu, nu/+, or other immunocompetent mice were inoculated with 10^6 yeast cells of *H. capsulatum* var. *capsulatum*,[49] the liver, spleen, and lymph nodes were affected in this order. In the liver of nu/+ or other immunocompetent mice, a few microabscesses with yeast cells appeared on the 2nd day. On the 4th day, the histo-

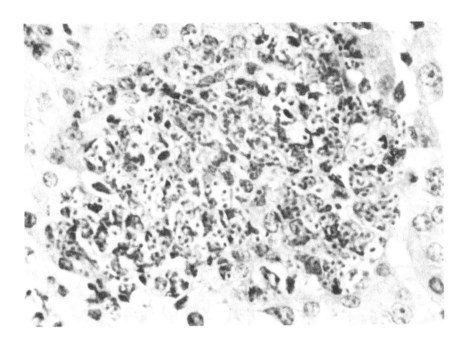

FIGURE 29. Granulomatous lesion with numerous yeast cells of *Histoplasma capsulatum* in the liver of a nu/nu mouse. (GMS-Hematoxylin-eosin.)

pathological features began to change into a granulomatous form. PMNs were scattered among mononuclear cells forming the granulomatous lesion, and yeast cells were found at its center. On the 6th day, the number of the granulomatous lesions with yeast cells increased, and on the 8th day, the number reached a peak. Thereafter, the number of lesions decreased and the yeast cells began to be destroyed in the granulomatous lesions. After the 11th day, it was very difficult to find yeast cells within the granulomatous lesions.

By comparison, the histopathological features in the livers of the nu/nu mice were nearly similar to those of the nu/+ mice during the early stages of infection. However, after the 8th day, there was a definite difference in the killing function between the granulomatous lesions of the nu/+ and nu/nu mice. Mononuclear cells forming granulomatous lesions could not kill the yeast cells, and as a result, the yeast cells continued to multiply in the granulomatous lesions in nu/nu mice. After the 18th day, most of the granulomatous lesions were occupied by numerous yeast cells (Figure 29).

The defense mechanisms of mice used against *H. capsulatum* var. *duboisii* were similar to those used against *H. capsulatum* var. *capsulatum*. Clinically, histoplasmosis duboisii[4] is not differentiated from histoplasmosis capsulati; it is essentially confined to the African continent and the point of difference from the latter is that the yeast cells of the former, in the lesion, are larger than those of the latter.[56] According to the author's data,[113] when nu/nu, nu/+, or other immunocompetent mice were inoculated intravenously with 10^6 yeast cells of *H. capsulatum* var. *duboisii*, the liver was the most severely affected, followed by the spleen and lymph nodes as in murine histoplasmosis capsulati. In the livers of the nu/+ mice, a granulomatous response occurred during the early stages of infection and the number of granulomatous lesions reached a peak on the 8th day and, thereafter, decreased as in histoplasmosis capsulati. Yeast cells in the granulomatous lesions began to be destroyed after that day. Granulomatous lesions also appeared in the liver of the nu/nu mice during the early stages of infection; however, the number of granulomatous lesions and yeast cells in them continued to increase with the advancement of infection, similar to the observations made when the mice were inoculated with *H. capsulatum* var. *capsulatum*. In murine histoplasmosis duboisii, yeast cells in the tissue were not necessarily larger than those in murine histoplasmosis capsulati.

Histoplasma farciminosum is the causative agent of a chronic disease of equines that generally involves the subcutaneous lymph nodes and lymphatics of the neck and legs.[4] The histopathological characteristics of the disease are a purulent and granulomatous inflammatory reaction. Such tissue responses are similar to those of histoplasmosis capsulati or histoplasmosis duboisii.

According to the author's data,[113] the virulence of *H. farciminosum* to nu/nu, nu/+, or immunocompetent mice was very low. When nu/nu or nu/+ mice were inoculated intravenously with 0.15 m*l* of a 4% cell suspension (wet weight), lesions appeared only in the liver and lungs of both groups of mice. When 0.15 m*l* of a 2% cell suspension was inoculated intravenously into mice, it was difficult to find lesions. The nu/+ mice infected with the former volume of the fungus had, in the livers, a few microabscesses with yeast cells on the 4th day. On the 6th day, mononuclear cells accumulated at the foci and most of the lesions became granulomatous, and on the 8th day, the number of lesions reached a peak, and yeast cells were found only in a few granulomatous lesions. Even though a considerable number of granulomatous lesions were observed in the liver tissue on the 11th and 14th days, only a few yeast cells, faintly stained by PAS, were found in a few granulomatous lesions. Comparatively, in the livers of the nu/nu mice, the primary histopathological feature during the early stages of infection was purulent. Different from the immunocompetent mice, the chief histopathological features did not change to granulomatous even on the 14th day; fungal cells, faintly stained by PAS, were found in a few microabscesses. On the 18th day, mononuclear cells accumulated at the foci and changed the histopathological features from pyogenic to granulomatous. Any mouse strains, including nu/nu mice, are not adequate as the experimental models of *H. farciminosum* infection.

4. Exophiala dermatitidis Infection

E. dermatitidis is one of the causative agents of chromomycosis. Even though the number of patients is not large, the fungus has been isolated occasionally from patients, especially in Japan.[57] The fungus is neurotropic and causes death in approximately one third of the patients. As a preliminary introduction, before describing experimental infections of nu/nu mice with the fungus, the virulence of 5 cultures of *E. dermatitidis* to ddY mice (male, weighing 19 to 22 g) is described. In 5 mice, each inoculated intravenously with 5×10^6 yeast cells of one (MM-7) of the cultures, 2 died on the 6th and 7th days. There were no dead mice in the other 4 groups of mice for a period of 30 days post-inoculation. In the dead mice, the brain was severely affected by the organism. A large number of PMNs accumulated around short hyphae and yeast cells and formed many microabscesses; however, a few microgranulomas with yeast cells were also observed in the brain tissue. Symptoms of central nervous system involvement (torticollis or abnormal movement) were observed in more than half of the mice inoculated intravenously with the five cultures, respectively. Thus, *E. dermatitidis* has a predilection for the mouse brain.

When nu/nu, nu/+, or other immunocompetent mice were inoculated intravenously with 10^6 yeast cells of *E. dermatitidis* (MM-7), the brain was the most favorable target organ of the fungus, followed by the kidneys and liver, in this order.[58] In the inoculated nu/+ mice, the colony-forming unit (CFU) from the brain reached a peak on the 4th day, then decreased gradually, and after the 40th day, no viable cells were recovered. By comparison, the CFU recovered from the brains of nu/nu mice was almost the same

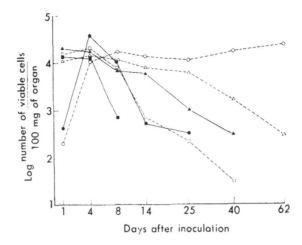

FIGURE 30.   Number of viable cells in the organs of nu/+ and nu/nu mice inoculated intravenously with 10⁶ yeast cells of *Exophiala dermatitidis* :○●, brain;△▲, kidney; and□■, liver. (Open symbols, nu/nu mice; closed symbols, nu/+ mice.) (From Nishimura, K. and Miyaji, M., *Mycopathologia*, 81, 11, 1983. With permission.)

as that recovered from the nu/+ mice until the 4th day. Thereafter, the number remained constant at that level until the 25th day, and then began to increase gradually (Figure 30).

In the brains of nu/nu mice, the chief histopathological findings during the early stages of infection were microabscesses (Figure 31-1), microgranulomas with short hyphae, and yeast cells. On the 6th day, mononuclear cells began to accumulate at the foci and the histopathological characteristics changed to a granulomatous type (Figure 31-2). At that time most of the fungal elements were yeast-like. Most of the granulomatous lesions were very small and were observed in a considerable number until the end of the experiment. Interestingly, the yeast cells were not killed in these granulomatous lesions, but their multiplication was suppressed. Thus, it seemed likely that a balance was maintained between the granulomatous lesions and the yeast cells. An acid-mucopolysaccharide, which was stained with alcian blue, was observed around the yeast cells after the 14th day (Figure 32). This substance might act as a buffer zone between the yeast cells and mononuclear cells forming the granulomatous lesions. Even though the chief histopathological features of the nu/nu mice were granulomatous after the 8th day, a few microabscesses and free hyphae were still observed until the end of the experiment (62nd day). The histopathological features in the brains of the nu/+ mice were almost the same as those in the nu/nu mice until the 10th day. Thereafter, the number of granulomatous lesions abruptly decreased and then the lesions began to disappear. No free fungal elements were observed.

In summary, the exertion of cell-mediated immunity was necessary for the killing of the fungus. However, in the brains of the nu/nu mice, mononuclear cells which formed granulomatous lesions suppressed fungus multiplication, even though they did not kill the fungus.

The virulence of other pathogenic black yeasts, such as *Exophiala jeanselmei*[59] and *E. spinifera*,[60] to ddY mice were as follows. When each ddY mouse (male, 19 to 21 g) was inoculated intravenously with 0.2 mℓ of a 1% cell suspension (wet weight) of *E. jeanselmei,* no mice died within 30 days of challenge. No mice grew weak from infection, nor did the central nervous syndrome appear. They gained weight steadily as if in

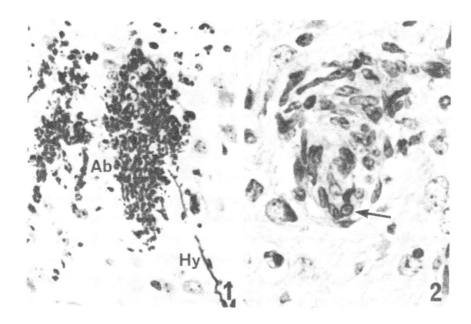

FIGURE 31.   Cellular responses in the brains of mice to *Exophiala dermatitidis* infection: (1) Abscess formation (Ab, abscess and Hy, hyphae) and (2) small granulomatous lesion with a yeast cell (arrow). (Periodic acid, Schiff.) (From Nishimura, K. and Miyaji, M., *Mycopathologia*, 77, 179, 1982. With permission.)

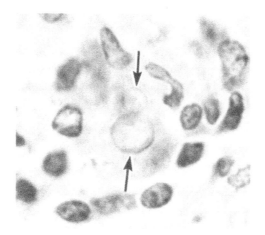

FIGURE 32.   Yeast-like cells (arrows) of *Exophiala dermatitidis* stained with alcian blue.

good health. The virulence of *E. jeanselmei* to ddY mice was much weaker than that of *E. dermatitidis*. Histopathologically, microfoci appeared mainly in the livers and lungs of the mice. The chief histopathological characteristics of the liver were Kupffer's cells engulfing yeast-like cells and a few purulent lesions during the early stages of infection. These lesions were scattered in the liver tissue. After the 7th day, the number of microgranulomatous lesions increased, and on the 26th day, most of the lesions disappeared. In the lung, a few microfoci of interstitial pneumonia with yeast-like cells sporadically appeared on the 2nd and 5th days. However, on the 7th day, almost all of the lesions had disappeared; there were no budding cells in the liver and lung lesions.

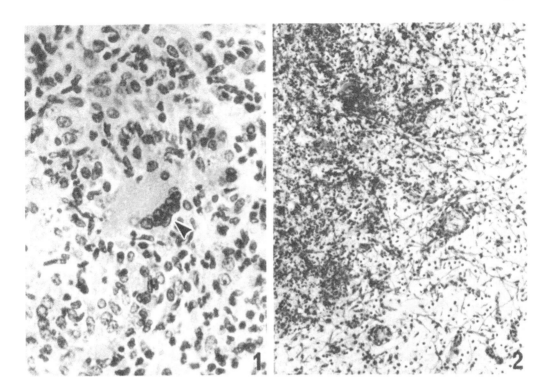

FIGURE 33.   Histopathological findings in the brains of nu/nu and nu/+ mice inoculated intravenously witl *Fonsecaea pedrosoi* : (1) Granulomatous lesion in the brain of a nu/+ mouse, and (2) purulent lesion in the brai of a nu/nu mouse. (Periodic acid, Schiff.)

The pathogenicity of *E. spinifera* was examined using ddY mice inoculated intravenously with 0.2 m*l* of a 1% cell suspension. Even though the fungus was recovered from the brain and kidneys for 30 days, the inflammatory reactions were mild. Microfoci were observed sporadically in the liver and kidneys during the early stages of infection. *E. spinifera* was not neurotropic like *E. dermatitidis.*

B. Fungal Infections in Which Both Polymorphonuclear Leukocytes and Activated Mononuclear Cells Play a Role in Host Defense

1. Fonsecaea pedrosoi Infection

F. pedrosoi is isolated most frequently as the causative agent of chromomycosis.[2-4] When nu/nu, nu/+, and ddY mice were inoculated intravenously with the fungus, the brain, kidneys, adrenal glands, and heart were affected most severely by the fungus, in that order.[61] First, the development of infection in the brains of nu/+ mice were described.[61] Spherical and short moniliform-inoculated hyphae clogged in the capillaries and germinated, breaking through the capillary walls within 4 days after inoculation. A few mononuclear cells migrated towards the hyphae. At almost the same time, PMNs were elicited to the hyphae and formed microabscesses. On the 6th day, the numbers of abscesses increased, their size enlarged, and after the 8th day, mononuclear cells began to accumulate at the foci, changing the histopathological character from pyogenic to granulomatous. On the 14th day, the granulomatous lesions (Figure 33-1) were completed, confining hyphae, preventing them from growing freely throughout the tissue, gradually destroying them. In the nu/nu mice, the histopathological features were similar to those of the nu/+ mice until the 6th day. Their histopathological feature was a marked suppurative reaction. However, an accumulation of mononuclear

Table 5

INVASIVENESS OF BRAIN BY *FONSECAEA PEDROSOI* AND CHIEF HISTOPATHOLOGICAL FEATURES

Inocula	Mice	Days after inoculation							
		1—4	6	8	10	14	18	21	25
0.02%	nu/nu	—	—	—	—	+1 P	—	+1 P	—
	nu/+	—	—	—	+1 PG	+2 PG	+2 PG	+3 PG	+3 PG
0.1%	nu/nu	—	—	—	+1 P	+2 P	+3 P	+3 P	+4 P
	nu/+	—	—	—	+1 PG	+2 PG	+2 G	+3 PGH	+2 G
0.5%	nu/nu	—	+1	+1 P	+2 P	+5 PH	/	/	/
	nu/+	—	+1	+1 PG	+1 PG	+3 PG	+4 PG	+3 PG	+3 PG

Note: +1: 0% < percentage $\leq$ 1%; +2: 1% < percentage $\leq$ 4%; +3: 4% < percentage $\leq$ 9%; +4: 9% < percentage $\leq$ 16%; +5: 16% < percentage; P: pyogenic response; G: granulomatous response, and H: hyphal growth without cellular response.

cells at the foci did not occur even 10 days after inoculation. As hyphae grew, the number of the abscesses increased and enlarged by consolidating with each other. After the 14th day, the hyphae broke the PMNs line and grew freely throughout the brain tissue without any observable cellular response (Figure 33-2).[61] However, when lymph node cells of nu/+ mice were adoptively transferred into nu/nu mice prior to challenge, granulomatous lesions were formed, confining the hyphae within them and preventing free growth of the hyphae.

According to another experiment on murine chromomycosis, a significant MIF response[62] developed 10 days after inoculation; this day closely coincided with the day when the granuloma formation began in the nu/+ mice. The titer of an agglutinating antibody, detected on the 21st day, however, was very low. Immunological tests in eight patients with chromomycosis caused by *F. pedrosoi* showed a positive delayed-skin reaction,[63] but the titer of the agglutinating antibody was very low, similar to that of the experimental murine chromomycosis model. These results appear to indicate that cell-mediated immunity might play a more important role in host defense against *F. pedrosoi* infection than does humoral immunity.

Table 5 presents a summary of the histopathological features observed in the brains of both groups of mice studied. In the nu/+ mice, the histopathological feature, until the 8th day, was a purulent reaction, and thereafter, a granulomatous one, while in the nu/nu mice, the feature was solely a purulent reaction or no cellular reaction at all. Thus, the defense in the nu/+ mice consists chiefly of two steps. In the first step, the phagocytosis by PMNs occurs, and in the second, cell-mediated immunity plays a role. The host defense exhibited by nu/nu mice, on the other hand, consists chiefly of only one step, phagocytosis by PMNs.

The difference in the susceptibility between the nu/+ and nu/nu mice to the fungus was due to the size of the fungal cell inoculum. When inoculated with 0.1 mℓ of the 0.02% cell suspension, the resistance of nu/nu mice was enhanced more than was that of the nu/+ mice. These data indicate that the killing function of PMNs in the nu/nu mice is more active than that in the nu/+ mice. According to Cutler[64] and Rogers et al.,[65] nu/nu mice were more resistant to *Candida albicans* infection than were nu/+ mice. Our results coincide with their results. There was little difference in the susceptibility to the *F. pedrosoi* infection between the nu/+ and nu/nu mice inoculated with 0.1 mℓ of the 0.1% cell suspension, thus indicating that the intensity of the defense mechanisms of the nu/+ mice, consisting of two steps, balances with that of the nu/nu mice, consisting of one step. When inoculated with 0.1 mℓ of the 0.5% cell suspension, the nu/nu mice lost their resistance to the fungal infection, and as a result, hyphae grew freely throughout the tissues of the brain and kidney.

2. Cladosporium trichoides Infection

C. trichoides, which is one of the pathogenic dematiaceous fungi, is neurotropic.[2-4] Interestingly, cerebral chromomycosis caused by the fungus is prone to occur in healthy men during their prime of life. Most of those infected die within a year. In animal experiments,[66-71] the brains were severely affected by this fungus.

All 10 nu/nu mice inoculated intravenously with 0.1 m*l* of a 1.0% cell suspension died by the 13th day, and those inoculated with 0.1 m*l* of a 0.1% cell suspension died by the 19th day.[72] There were no dead nu/nu mice, inoculated with 0.1 m*l* of a 0.01% cell suspension, until the 30th day. Comparatively, all 10 nu/+ mice had died by the 19th day when they were inoculated intravenously with 0.1 m*l* of the 1.0% cell suspension. However, no nu/+ mice which had been inoculated with either of the 0.1 and 0.01% cell suspension died during the experimental period (30 days).

There were no characteristic pathological findings in the nu/+ and nu/nu mice inoculated with the 0.01% cell suspension. When inoculated with the 1.0% cell suspension, the brain, in both nu/nu and nu/+ mice, was the primary target organ of the fungus, followed by the kidney. When inoculated with the 0.1% cell suspension, brain lesions were observed only in the nu/nu mice. The histopathological findings indicated that the death of the mice was caused by the fungus invasion of the brain. Even though lesions were observed in the liver, spleen, lung, heart, and lymph nodes, the degree of infection was mild compared to that in the brain and kidneys. In nu/+ mice inoculated with the 1.0% cell suspension, there were no characteristic pathologic findings on the first day. On the 4th day, a few short, slender hyphae appeared in the brain tissue eroding through the capillary walls. A few mononuclear cells were attached to the hyphae. Immediately, PMNs were elicited to the hyphae and several microabscesses were formed. Thereafter, the lesions continued to enlarge until the 19th day; at that time, hyphae grew freely throughout the brain tissue. In nu/nu mice infected with the 1.0% cell suspension, a few short hyphae in conjunction with a few mononuclear cells appeared on the 6th day. The slender hyphae continued to grow rapidly, and on the 11th and 13th days, numerous hyphae without an associated cellular response were observed throughout the brain tissue (Figure 34). In nu/nu mice infected with the 0.1% cell suspension, numerous microabscesses with hyphae were found in the sections after the 14th day. Interestingly, different from *F. pedrosoi* infection,[61] an accumulation of mononuclear cells did not occur at the foci in the brain of the nu/+ mice even after the 19th day. Perhaps, the hyphal growth of the fungus was so rapid that mononuclear cells had no time to accumulate at the lesions and form granulomatous lesions.

3. Blastomyces dermatitidis Infection and Paracoccidioides brasiliensis Infection

B. dermatitidis and *P. brasiliensis* are the causative agents of blastomycosis and paracoccidioidomycosis, respectively. The former occurs frequently in the area along the Mississippi River valley and the latter, in the South American continent.[2-4] Usually, the primary lesion occurs in the lung after inhalation of conidia of either of the fungi. The histopathological characteristic in both diseases is a mixed reaction of purulent and granulomatous inflammation.

In nu/nu mice inoculated intravenously with 0.2 m*l* of a 1% cell suspension of *B. dermatitidis* or with 0.15 m*l* of a 2% cell suspension of *P. brasiliensis,* a marked purulent inflammatory reaction was induced against each fungus during the early stages of infection, and relatively large abscesses were formed in the brain tissue. Even though a number of yeast cells were killed in the abscesses, a considerable number of

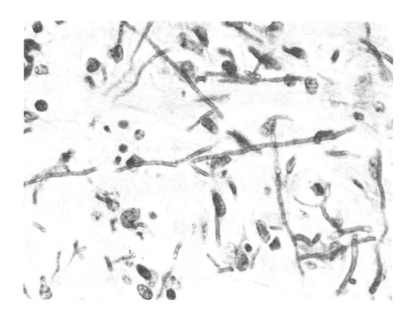

FIGURE 34.   Hyphal growth of *Cladosporium trichoides* in the brain of a nu/nu
mouse. (Periodic acid, Schiff.)

fungal cells survived. On the 8th day, mononuclear cells began to accumulate at the
foci, and on the 12th or 18th day, more than half of the lesions of the nu/nu mice
inoculated with *B. dermatitidis* or *P. brasiliensis* became granulomatous. However,
different from the granuloma formed in the brain of the nu/+ mice, the mononuclear
cells which formed the granulomatous lesion could not kill the yeast cells, so that the
yeasts continued to multiply in the granulomatous lesions (Figures 35 and 36).[73]

4. Candida albicans Infection

C. albicans is one of the etiologic agents of opportunistic fungal infections. It is one
of the members of the normal flora of the skin and of the alimentary canal. When the
resistance of a host weakens for some reason, it invades the tissue. The virulence of *C.
albicans* to mice varies; usually, in experimental murine candidiasis, 10^5 to 5×10^6 yeast
cells are inoculated intravenously, intraperitoneally, or subcutaneously. To date, there
have been published studies[64,65,74-76] on experimental murine candidiasis in nu/nu and
nu/+ mice. In nu/nu, nu/+, or other immunocompetent mice inoculated intravenously
with the fungus, the kidney, brain, and liver were affected, in that order. According to
Cutler[64] and Rogers et al.,[65] the resistance of nu/nu mice to *C. albicans* infection was
stronger than that of nu/+ mice. According to Tabeta et al.,[74] when nu/nu and nu/+
mice were inoculated intravenously with 5×10^5 yeast cells, the nu/+ mice showed a
shorter survival period. However, when these mice were inoculated intravenously with
3×10^6 yeast cells, there was no difference in survival in either group of mice. Lee and
Balish[75] also described that the resistance of conventionalized (ex-flora-defined) nu/nu
mice to *C. albicans* was stronger than that of control mice. However, when the nu/nu
mice raised under germfree or flora-defined conditions were used, their susceptibility
increased more than that of control mice. Furthermore, they[76] reported a role played
by macrophages in the host defense to *C. albicans* infection using nu/nu and nu/+
mice. They inoculated silica intravenously into these mice to assess the role of macrophages. According to the authors' results,[113] when nu/nu and nu/+ mice were inoculated intravenously with 5×10^5 yeast cells, histopathological changes were found

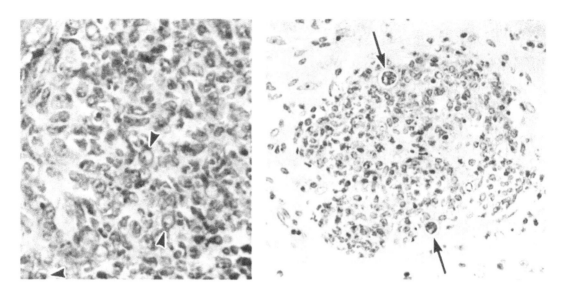

FIGURE 35   FIGURE 36

FIGURE 35.   Granulomatous lesion in the brain of a nu/nu mouse inoculated intravenously with *Blastomyces dermatitidis*. Yeast cells (arrow heads) multiply in the lesion. (Periodic acid, Schiff.)

FIGURE 36.   Granulomatous lesion with spherical cells (arrows) in the brain of a nu/nu mouse inoculated intravenously with *Paracoccidioides brasiliensis*. (Periodic acid, Schiff.) (From Miyaji, M. and Nishimura, K., *Mycopathologia*, 82, 136, 1983. With permission.)

chiefly in the brain and kidneys. In the brain of both groups of mice, abscesses with fungal elements were formed by the 4th day. However, the purulent inflammatory reactions in the nu/nu mice were milder than those of the nu/+ mice, and on the 6th day, there were no reactions found in the nu/nu mice. In the nu/+ mice, a few granulatous lesions were found on the 6th day, but thereafter, there were no histopathological findings.

On the other hand, the kidneys of the nu/+ mice were affected more severely than were those of the nu/nu mice by the 8th day, but thereafter, the latter was affected more severely than the former (Figure 37). These results seem to indicate that phagocytosis by PMNs play an important role in host defense to *C. albicans* during the early stages of infection and cell-mediated immunity, during the later stages of infection. Thus, the host defense to *C. albicans,* in immunocompetent mice, consists chiefly of two steps as in *F. pedrosoi, B. dermatitidis,* or *P. brasiliensis* infection. The defense in the nu/nu mice consists chiefly of one step, phagocytosis by PMNs.

5. Coccidioides immitis Infection

C. immitis is the causative agent of coccidioidomycosis which is limited to the southwestern part of the United States, Mexico, and Central and South America. The disease usually occurs after inhalation of the arthroconidia of *C. immitis,* which change to spherules packed with numerous endospores. Histopathologically, the tissue responses closely relate to the developmental stages of spherules. The histopathological features against young spherules are a marked purulent-inflammatory reaction and those against mature spherules, a granulomatous inflammatory one.[2-4]

The course of *C. immitis* infection in Swiss white mice were as follows. When these

FIGURE 37.   *Candida albicans* (arrows) in the pelvis of the kidney of a nu/nu mouse. (Periodic acid, Schiff.)

mice were inoculated intravenously with 5×10^4 arthroconidia, the liver, lung, and spleen were affected severely, in that order.[77] The histopathological feature in the liver was a marked purulent-inflammatory reaction during the early stages of infection. The abscesses with spherules were surrounded by a zone of the liquefied tissue 6 days after inoculation. The nuclei of the liver cells were rarely recognizable within this zone (Figure 38-1). On the 8th day, relatively large lesions consisting of three zones appeared (Figure 38-2). The central necrotic area consisted of PMNs and their debris; the intermediate area consisted of a zone of the liquefied tissue. The outer area reflected a granulomatous tissue consisting of mononuclear cells; a few fungal elements were observed in the necrotic area of the inner zone. After the 10th day, numerous mononuclear cells accumulated around the necrotic areas and formed granulomatous lesions. On the 14th day, ruptured spherules, which were attacked by PMNs, were observed in the central necrotic area surrounded by the granulomatous tissue. The influx of PMNs toward the inside of the spherules was observed. On the 18th day, the chief histopathological features were a granulomatous response and small spherules with or without purulent reaction in the granulomatous lesions.

To date, there have been three papers dealing with nu/nu mice infected with *C. immitis*. In 1977, Beaman et al.[78] reported that nu/nu mice were more susceptible to *C. immitis* infection than were nu/+ mice. However, since their conclusion was based on the results of mortality, their description concerning the histology was poor. There were meager cellular responses to spherules which grew abundantly in the tissue of the nu/nu mice. According to the authors,[20] groups of nu/nu and nu/+ mice (consisting of 7 mice each) inoculated intravenously with 3.3×10^5 arthroconidia, died on the 6th and 7th days, respectively. As the purpose of the experiment was to observe the parasitic cycle of the fungus, their description concerning the histology was also poor. The liver, lung, and spleen were affected severely, in that order, and cellular responses to spherules were mild in both groups of mice.

Recently, Clemons et al.[79] studied quantitatively and histologically the course of *C. immitis* in beige mice, nu/nu mice, and their respective controls. As for histology, there were no characteristic findings in the organs of these mice by the 14th day. Thereafter, there was a definite difference in the cellular responses between the nu/nu and nu/+

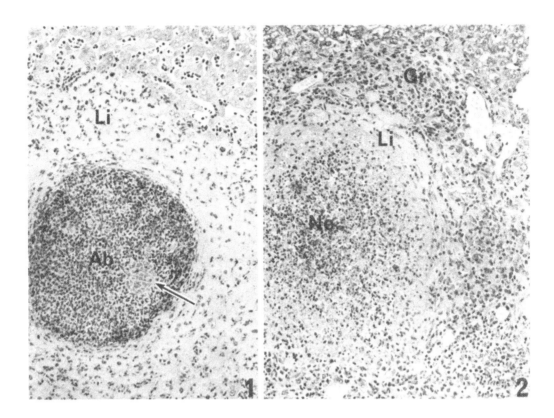

FIGURE 38.   Histopathological findings of the livers of mice inoculated intravenously with *Coccidioides immitis*: (1) abscess (Ab) with broken spherules (arrow) is surrounded by a zone of the liquefied tissue (Li) 6 days after inoculation; (2) lesion consisting of 3 zones, a necrotic area (Ne) is observed at the center and surrounded by the middle zone of the liquefied tissue (Li) while the outermost zone consists of a granulomatous tissue (Gr). Eight days after inoculation. (Periodic acid, Schiff.)

mice. The chief histological feature in the nu/nu mice was suppurative and that in the nu/+ ones was granulomatous. In addition, the extent of the inflammatory reactions in the former mice was less than that in the latter, and the number of spherules observed in the nu/nu mice was often greater than that in the nu/+ mice. From these and other results, they concluded that the host defense against *C. immitis* infection consists of two steps: the phagocytosis by PMNs during the early stages of infection and the killing by the activated macrophages during the later ones.

C. Fungal Infections in Which Polymorphonuclear Leukocytes Play a Leading Role in Host Defense

1. Aspergillus fumigatus Infection

   A. fumigatus is the most frequent causative agent of aspergillosis; like *C. albicans*, it is one of the most common causative agents of opportunistic fungal infections. The range of the virulence of *A. fumigatus* for mice is wide. Usually, in experimental murine aspergillosis, *A. fumigatus* has a predilection for the kidney; however, a few cultures, in particular, have a predilection for the brain of the mouse. According to the authors[80,81] 3 of 25 cultures of *A. fumigatus* had a predilection for the brain of mouse. Naturally, the virulence of these 3 cultures in mice was strong. When ddY mice (male, 20 to 22 g) were inoculated intravenously with 5×10^6 conidia of the *Aspergillus*, all of the mice died within 1 week. Macroscopically, the brains were markedly swollen more

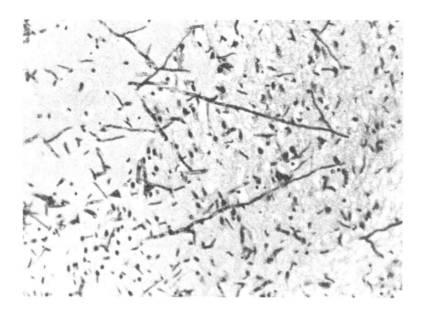

FIGURE 39.   Proliferation of *Aspergillus fumigatus* in the brain of a mouse. (Periodic acid, Schiff.)

than 40 hr after inoculation. The cranial sutures were loosened, and hemorrhagic foci of various sizes were observed on the surfaces of the brains. Microscopically, a few hyphae appeared in the cerebral tissue 16 hr after inoculation. Several necrotic foci with hyphae, cellular infiltration (PMNs), and hemorrhage were observed in the tissue 28 hr after inoculation. Abundant hyphal growth with or without purulent reactions were observed in the brain tissue more than 40 hr after inoculation (Figure 39). These three cultures were high in proteolytic activity when cultured on brain-heart infusion agar.[82]

According to Arai et al.,[83] there was no difference in resistance to *A. fumigatus* infection between nu/nu and nu/+ mice. When 10 nu/nu and 10 nu/+ mice were inoculated intravenously with 10^6 conidia of one of the 3 cultures mentioned above and observed for 30 days, 2 of the former group and 1 of the latter group died. There was no difference in histopathology between nu/nu and nu/+ mice. Histopathologically, a purulent inflammatory reaction was apparent until the 3rd day, and on the 5th day, mononuclear cells began to accumulate at the purulent lesions, changing the histopathological character to a granulomatous reaction. Usually, in murine aspergillosis, PMNs play a leading role in host defense; additionally, in experimental murine zygomycosis (mucormycosis), PMNs play a leading role in host defense as well. According to Corbel and Eades,[84] there were no differences in resistance and histopathology between nu/nu and nu/+ mice.

D. Conclusions

The pathogenic fungi may be classified into three categories according to the results obtained from nu/nu, nu/+, and other immunocompetent mice. *S. schenckii, H. capsulatum, C. neoformans,* and *E. dermatitidis* are classified into the first category. The mononuclear cells activated by T-lymphocytes play a leading role in host defense against these fungal infections. The second category includes *F. pedrosoi, B. dermatitidis, P. brasiliensis, C. trichoides,* and *C. albicans.* The host defense against these causative agents consists of two steps: the phagocytosis by PMNs during the early stages of infection and the killing by the activated macrophages during the later stages of infection. Possibly *C. immitis* belongs in this category. Fungi classified into the 3rd category are *A. fumigatus*, zygomycetes, *Fusarium*, *Pseudallescheria boydii*, etc. In mice infected with these fungi, PMNs play a leading role in host defense.

V. EFFECT OF SPLENECTOMY UPON GRANULOMA FORMATION AND KILLING FUNCTION OF GRANULOMA AGAINST FUNGAL INFECTIONS

In the previous section we described histopathologically the course of various fungal infections and the defensive role of granuloma against them. As described in the section, the granuloma is formed without the interaction of cell-mediated immunity, which, however, is indispensable for the killing of fungi in the granuloma. Therefore, when functional cell-mediated immunity is abrogated by the administration of cortisone, immunosuppressive drugs, radiation, or by suffering from such exhausting diseases as leukemia, Hodgkin's disease and other malignant tumors, metabolic and autoimmune diseases, etc.,[85] the killing functions against fungi of the granuloma weakens. Usually, in these cases, other host defenses, such as phagocytosis by PMNs, humoral immunity, etc., are injured at the same time, so that the fungi can invade the deep tissues more easily.

As is well-known, cell-mediated immunity is controlled by T-lymphocytes from the thymus; the spleen also influences the functions of cell-mediated immunity[86] because the white pulp contains T-lymphocytes in the periarteriolar sheaths. Therefore, there is the possibility that the spleen might play an important role in regulating the killing function of the granuloma against invading fungi. However, there have been no reports focusing on this subject in the field of medical mycology.

In this section, the role the spleen plays in host defense against *S. schenckii* and *C. neoformans* infections is described.

A. Effect of Splenectomy upon Microorganism Infections

The spleen plays an important defensive role against microorganism infections. In particular, in the field of pediatrics, splenectomized patients have been affected severely by bacterial infections.[87-89] Experimentally, it has been confirmed that the susceptibility of splenectomized animals to bacterial or protozoal infections increases greatly when compared to that of control animals.[90-96] However, interestingly, adverse cases have been reported. According to Skamene and Chayasirisobhon,[97] splenectomized mice survived an infection with a dose of *Listeria monocytogenes* 100 times larger than did intact or sham-operated mice. In inbred B10LP mice infected with *Plasmodium berghei*, the average survival time of splenectomized mice was three times longer than that of intact mice.[98]

B. Splenomegaly in Murine Mycoses

The authors have carried out many animal experiments using various species of pathogenic fungi. During the experiments they noticed splenomegaly in mice inoculated intravenously with *S. schenckii*,[46] *H. capsulatum*,[49] and *C. neoformans*.[40]

Usually the average weight of the spleen of mice is less than 150 mg regardless of strain, body weight, or age of mice. When mice were inoculated intravenously with either of the 3 fungal species, the average spleen weight was more than 250 mg, and in some cases, it was more than 500 mg 10 days after inoculation. These results appear to indicate that the spleen might play some role in host defense against these fungal infections. In bacterial or protozoal infections, the splenomegaly is caused by an increase of B- and T-lymphocytes.[99] Interestingly, in these fungal infections, splenomegaly occurred even in nu/nu mice, as well as in immunocompetent mice. Histopathologically, the splenomegaly was caused by the increase in number of mononuclear cells. As described in the previous section, mononuclear cells activated by interaction with cell-mediated immunity play a leading defensive role against these fungal infections. Therefore, it is interesting to know how the spleen relates to the granuloma formation and the killing function of the granuloma for these fungi.

C. Splenectomy

The authors used male ddY and BALB/c mice, weighing 19 to 24 g, in the following experiment.[100] A mouse was anesthetized with diethyl-ether according to the procedure described for the agar implantation method. The anesthetized mouse was laid on its face on a tray covered with a sheet of aluminum foil. A small reversed beaker whose bottom was covered with absorbent cotton soaked with diethyl ether was put on its head and the left side of the back was opened by a small incision after disinfecting the operation area with 70% alcohol. The spleen was carefully taken out from the abdominal cavity and removed, after both Arteria linealis and Vena linealis were ligated with thread and the wound was closed. In the experiment using operated animals, at least two kinds of controls are demanded. One is the experiment using intact mice and the other is one using mice with sham-operation. In this experiment, the sham-operation was carried out as follows. The spleen was taken out from the abdominal cavity according to the procedure described above, returned immediately into the abdominal cavity, and the wound was closed. If possible, it is best to administer antibiotics into the mouse after the operation. Even though approximately half of the splenectomized mice lost consciousness, they recovered within a few hours. The splenectomized mice were maintained in a glove box as were the nu/nu mice. Challenge with fungal cells was done 7 days after the operation.

D. Resistance of Splenectomized Mice to *Sporothrix schenckii* Infection[100]

Male ddY mice (19 to 23 g) were divided into three groups consisting of splenectomized, sham-operated and intact mice, respectively. Each mouse was inoculated intravenously with 2×10^6 yeast cells of *S. schenckii* 7 days after surgery, and the mice were killed at various intervals for 30 days.

The growth rate of the mice was conspicuously influenced by splenectomy, but the splenectomized mice were hardly influenced by inoculation of *S. schenckii*. On the other hand, the growth rate of the sham-operated mice slowed down after inoculation of the fungus as shown in Figure 40.

There was a distinct difference in the spleen weights between mice which had neither surgery nor inoculation of *S. schenckii* and the sham-operated or intact mice with inoculation. The average weight of the spleen of the 10 sham-operated and 10 intact mice infected with *S. schenckii* was 276 and 269 mg, respectively, 9 days after inoculation, while that of the 10 intact mice without infection was 140 mg.

When mice were inoculated intravenously with *S. schenckii,* the liver was affected most severely, thus the authors focused on the liver. In the liver of the intact and sham-operated mice 2 days after inoculation, there were a few microfoci consisting mainly of PMNs, with or without yeast cells. Many microfoci appeared on the 5th day. The chief histopathological feature was a purulent-inflammatory reaction. On the 8th day, mononuclear cells accumulated at the foci, changing the histopathological features from pyogenic to granulomatous. On the 10th day, most of the foci became granulomatous lesions with yeast cells. At that time, the number of granulomatous lesions reached a peak. On the 14th day, the number decreased abruptly and yeast cells in

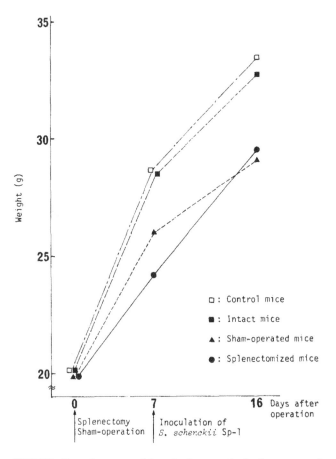

FIGURE 40. Average weight of splenectomized, sham-operated, and intact mice inoculated intravenously with 2×10^6 yeast cells of *Sporothrix schenckii*. (From Miyaji, M. and Nishimura K., *Mycopathologia*, 96, 145, 1986. With permission.)

them were destroyed. In the splenectomized mice, a few microabscesses with or without yeast cells appeared on the 2nd day, similar to the intact and sham-operated mice. On the 5th day, approximately half of the foci changed to granulomatous lesions. Unlike the course of infection of the intact or sham-operated mice, the number of foci began to decrease after the 5th day, and most of the lesions became granulomatous on the 8th day. Different from the histopathological findings of the intact or sham-operated mice, a considerable number of lymphocytes accumulated at the periphery of the granulomatous lesions in the splenectomized mice from 8 to 14 days after inoculation.

The number of yeast cells in the liver section reached a peak on the 10th day in the intact and sham-operated mice infected with *S. schenckii,* and thereafter decreased abruptly. However, in the splenectomized mice, the number of yeast cells reached a peak on the 5th day and, thereafter, it decreased (Table 6). Thus, in the splenectomized mice, the yeast cells began to be destroyed approximately 5 days earlier than in the intact and sham-operated mice. Table 7 presents the number of yeast cells observed in liver sections of 30 mice killed 9 days after inoculation. There was a definite difference between the splenectomized and control mice (P < 0.01).

As shown in Table 8, there was a definite difference in the number of CFUs of the liver between the splenectomized and the control mice killed 9 days after inoculation (P < 0.01). The killing function of splenectomized mice for the yeast cells of *S. schenckii* was greater than that of the intact or sham-operated mice.

Table 6

NUMBER OF YEAST CELLS IN 40 mm^2 OF LIVER SECTIONS OF
SPLENECTOMIZED, SHAM-OPERATED, AND INTACT ddY MICE
INOCULATED INTRAVENOUSLY WITH 2 × 10^6 YEAST CELLS OF
SPOROTHRIX SCHENCKII Sp-1

	Days after inoculation									
	2	5	8	9	10	14	16	21	26	30
Splenectomized mice										
Number of yeast cells	19	123	90	82	60	33	28	8	8	5
Parasitic form	C	C > R	C = R	C = R	C < R	C < R	C < R	C < R	R	R
Sham-operated mice										
Number of yeast cells	12	198	208	235	240	57	18	12	12	5
Parasitic form	C	C > R	C > R	C > R	C = R	C < R	C < R	C < R	R	R
Intact mice										
Number of yeast cells	14	118	238	201	268	42	15	16	10	7
Parasitic form	C	C > R	C > R	C > R	C = R	C < R	C < R	C < R	R	R

Note: Mice were splenectomized or sham-operated 7 days before inoculation. C: Cigar shaped; R: round or oval.

Table 7

NUMBER OF YEAST CELLS IN 40 mm^2 OF
LIVER SECTIONS OF SPLENECTOMIZED,
SHAM-OPERATED, AND INTACT MICE
INOCULATED WITH 2 × 10^6 YEAST CELLS
OF *SPOROTHRIX SCHENCKII* Sp-1

Mouse number	Splenectomized mice	Sham-operated mice	Intact mice
1	83	225	189
2	81	232	190
3	70	230	228
4	81	252	190
5	78	259	196
6	89	246	211
7	87	234	210
8	81	219	186
9	86	216	210
10	82	239	203
Average	82	235	201

Note: $p < 0.01$, the distribution of t by Student's test. Mice were sacrificed 9 days after inoculation.

E. Resistance of Splenectomized Mice to *Cryptococcus neoformans* Infection

Usually, when mice are inoculated intravenously with *C. neoformans*, the brain and liver are the primary target organs. As described in Section IV, the histopathological characteristics in murine cryptococcosis are cystic in the brain and granulomatous in the liver. Therefore, in the brain parenchyma, mice have no defensive measures against *C. neoformans* infection.

The effects of splenectomy on murine cryptococcosis were as follows. Male BALB/ c mice, weighing 20 to 24 g, were divided into 3 groups, consisting of splenectomized,

Table 8

COLONY-FORMING UNIT IN 100 mg OF
LIVER OF SPLENECTOMIZED, SHAM-
OPERATED, AND INTACT MICE
INOCULATED WITH 2×10^6 YEAST CELLS
OF *SPOROTHRIX SCHENCKII* Sp-1[a]

Mouse number	Splenectomized mice	Sham-operated mice	Intact mice
1	2.0×10^5	7.5×10^6	3.5×10^6
2	1.7×10^6	9.1×10^6	8.4×10^6
3	0.9×10^5	3.6×10^6	9.6×10^6
4	1.3×10^5	2.6×10^6	1.4×10^6
5	1.4×10^5	2.0×10^6	5.0×10^6
6	6.0×10^5	3.9×10^6	2.1×10^6
7	4.9×10^6	9.7×10^6	4.7×10^6
8	4.0×10^5	1.8×10^6	7.4×10^6
9	3.5×10^5	7.9×10^6	5.3×10^6
10	8.0×10^5	7.6×10^6	3.4×10^6
Average	9.3×10^5	5.6×10^6	5.1×10^6

Note: $P < 0.01$, the distribution of t by Student's test. Mice
were sacrificed 9 days after inoculation.

sham-operated, and intact mice. Each mouse was inoculated intravenously with 10^6 yeast cells of *C. neoformans* 7 days after the operation. There were no lesions in the brain of the splenectomized, sham-operated, and intact mice 2 days after inoculation. Five days after inoculation, cystic lesions were observed in the brains of the three groups of mice. Thereafter, the cystic lesions with cryptococci continued to increase regardless of the mouse groups. There were no other histopathological findings in the brains of the three groups of mice, and cellular responses did not occur in the brain parenchyma by the end of the experiment (30 days). There was no difference in the number of cystic lesions or in the area of the cystic lesions between the splenectomized and control (sham-operated and intact) mice.

The course of infection in the liver tissue was as follows. In the control mice, a few microabscesses were found in the liver 2 days after inoculation. The number of micro-foci increased until the 14th day after inoculation, and thereafter decreased abruptly. The chief histopathological feature was purulent during the early stages of infection and, after the 8th day, changed from pyogenic to granulomatous. Seventeen days after inoculation, most of the yeast cells in the granulomatous lesions were destroyed. In the splenectomized mice, a few microfoci were found on the 5th day, and the number of foci continued to increase until the 20th day after inoculation. Thereafter, the number of foci decreased. The histopathological features were similar to those of the controls; however, even on the 20th day, yeast cells were found in the granulomatous lesions without being destroyed. The yeast cells were destroyed in the granulomatous lesions 25 days after inoculation. There was a definite difference between the splenectomized mice and control ones in the numbers of both granulomatous lesions and yeast cells 20 days after inoculation (Table 9), even though there was no difference in the numbers 10 and 30 days after inoculation. Thus, the killing functions in the liver of the splenectomized mice for *C. neoformans* were weakened after splenectomy.

In conclusion, there was a delay of approximately 5 days in the killing functions of the granuloma in the liver of the splenectomized mice.

Table 9
NUMBER OF GRANULOMATOUS LESIONS AND
YEAST CELLS IN 30 MM² OF LIVER SECTIONS
OF MICE INOCULATED INTRAVENOUSLY
WITH 10⁶ CRYPTOCOCCI

Mouse group	Number of granulomatous lesions	Number of yeast cells
Splenectomized mice	50.7 ± 6.76	44.3 ± 4.14
Sham-operated mice	22.6 ± 1.95	22.6 ± 2.33
Intact mice	21.8 ± 3.99	20.9 ± 4.48

Note: Each group consisted of 9 mice. P < 0.01. Mice were sacrificed 20 days after inoculation.

VI. OPPORTUNISTIC FUNGAL INFECTIONS IN ACUTE LYMPHOCYTIC LEUKEMIA

Opportunistic fungal infections frequently occur in patients with malignant tumors during the terminal stages of the disease.[85] In particular, disseminated fungal infections are increasing in frequency in the patients with acute lymphocytic leukemia. These patients are treated with broad-spectrum chemotherapeutic drugs, which further accelerate fungal infections. According to Abe et al.,[101] 36 (66.7%) out of 54 cases with acute lymphocytic leukemia were infected with fungi, among which *Candida* was isolated from 26 out of the 36 cases as the causative agent, followed by *Aspergillus* (18 cases), zygomycetes (6 cases), and *C. neoformans* (1 case) (since dual infection occurred in some cases, the total number was more than 36). Therefore, it is significant and appropriate to use leukemic mice in the study of opportunistic fungal infections.

To date there have been few reports on experimental fungal infections in tumor-bearing mice. Uetsuka et al.,[102] studied the protective effect of a protein-bound polysaccharide substance against candidiasis using tumor(sarcoma 180)-bearing mice. According to Mardon et al.,[103] the susceptibility of Lewis lung carcinoma-bearing mice to *C. albicans* was greater than that of noncancerous animals. Johnson et al.[104] reported that disseminated candidiasis occurred in mice challenged with L1210 leukemic cells more frequently than in control ones. Abe et al.[105] studied in detail a relationship between leukemia and fungal infections using AKR/J leukemic mice as a suitable animal model for the study of human lymphocytic leukemia. The leukemic mice have to be maintained under a specific pathogen-free condition. Usually, they are maintained in a glove box, and supplied with sterilized cages, bedding, food, and water. A characteristic of the leukemic mice is that their thymus swells conspicuously more than 6 months after birth. To know whether or not the mice are suffering from leukemia, the following symptoms are adopted as criteria:[105] (1) conspicuous loss of body weight, (2) acceleration of breath, and (3) bristly hair. Furthermore, examination of their blood profile and the number of blood cells is required for the confirmation of leukemia. However, it is difficult to sample blood from the mice suffering from leukemia because they die readily from suffocation if handled roughly because of the remarkable swelling of the thymus. The blood is sampled from the venous plexus in the orbit of an eye with a hematocrit tube. Usually, blood can be collected from a mouse only two times, that is just before the start of experiment and just before death.

When the leukemic mice were inoculated intravenously with 10⁶ blastoconidia of *C.*

albicans, the average duration of their survival was 10.4 days, while for control mice (ICR mice), only one mouse died 16 days after inoculation.[105] When the leukemic mice were administered steroid hormone (hydrocortisone acetate, 1 mg/day) for 3 consecutive days just after inoculation of *C. albicans*, all died within 9 days after challenge. Marked peripheral lymphocytopenia occurred in the mice which were administered steroid hormone. Histopathologically, in the leukemic mice, the kidney was affected most severely, followed by the brain. In the mice which died during the early stages of infection, no cellular responses occurred at clumps of *C. albicans*, but in those which died after the 6th day, a moderate cellular infiltrate, consisting mainly of PMNs, was induced to the fungus.[105]

When 10^6 conidia of *A. fumigatus* were inoculated intravenously into the leukemic mice,[106] the average length of survival was 8.5 days, even though all of the control mice survived until the end of the experiment (30 days). Histopathologically, the kidneys were affected most severely, followed by the brain and heart. Hyphae grew freely or with mild cellular responses. When leukemic mice were administered hydrocortisone acetate (1 mg/day) for 3 consecutive days just after inoculation of *A. fumigatus*, all died within 5 days. According to Abe et al.[106], the number of lymphocytes in the peripheral blood of ICR mice conspicuously decreased 24 hr after the administration of hydrocortisone acetate and did not increase again even 2 weeks after administration.

VII. SUMMARY OF EXPERIMENTAL FUNGAL INFECTIONS

As described in Section IV, the host defense to fungal infections depends chiefly on killing by PMNs and mononuclear cells forming granuloma. In particular, the killing by the latter related closely to the exertion of cell-mediated immunity. Which of them plays a primary role in host defense depends on fungus species.

We can analyze the particular host defense against various fungal infections by suppressing or accelerating the functions of PMNs, mononuclear cells, cell-mediated immunity, etc. For example, the number and functions of PMNs are decreased and suppressed, respectively, by radiation, or an administration of nitrogen mustard. In this case, as the sessile macrophages of the reticuloendothelial system usually remain intact, it is possible to analyze the role PMNs play in the host defense. When carrageenan, dextran sulfate, or silica is administered intravenously, the functions of monocytes in the peripheral blood and sessile macrophages decrease selectively due to the saturation of phagocytic capacity; therefore, by administration of these substances, we can analyze the defensive role of mononuclear cells. In the study of the defensive role of T-lymphocytes, we can use nu/nu or thymectomized mice. In addition, subsets of T-lymphocytes are injured by the administration of cyclophosphamide. In rats, radiated with X-rays after thymectomy and then adoptively transferred bone-marrow from the same strain of rat, the functions of T-lymphocytes do not recover, but those of PMNs and B-lymphocytes do.

Usually, mycoses are prone to occur in the patients in whom any of the defense mechanisms are lowered, that is to say, in compromised hosts. When the concentration of some nutrient essential for fungal growth increases in compromised patients, fungal growth further accelerates. One example of such a nutrient is iron which is an essential mineral for fungal as well as bacterial growth. In the vessels and tissues, some of the causative agents of opportunistic fungal infections produce iron-transport cofactors (siderophore, siderochrome) which bind with iron and form a chelate, and the compound is absorbed by the fungi. These iron-transport cofactors compete with host transferrin in binding iron. In immunocompetent persons, their transferrin is not saturated (at a saturation rate of 20 to 30%), it easily binds with iron, and as a result, fungi cannot compete sufficiently for the available iron. Sometimes, in patients with hepatitis, iron is released from the liver cells[107] and absorbed into the serum, and in those with leukemia, lymphoma, thalassemia, etc., it is released from erythrocytes.[107] In these cases the *unbound-iron binding capacity* (UIBC) of host transferrin is reduced, and fungal growth is accelerated by use of excess iron.

To date, there have been few reports dealing with this subject in vivo in the field of medical mycology. Abe et al. studied the effects of iron overload on *C. albicans*[108] and *A. fumigatus*[109] infections. They used ICR mice and administered intravenously chondroitin sulfate colloidal iron for 3 consecutive days (40 or 60 mg of iron per kg of weight per day) to experimental mice, and concluded that there was a marked difference in mortality between the mice with iron overload and the control ones infected intravenously with either of the two organisms. The high mortality rate in the mice with iron overload is due to both increased serum iron (decreased UIBC) and decreased phagocytic activity of the sessile macrophages, whose phagocytic capacity is saturated by the engulfment of colloidal iron. In either experiment, enhanced proliferation of *C. albicans* or *A. fumigatus* was observed in the kidneys, brain, and heart, in that order.

It is well-known that a fulminant infection with zygomycetes occurs sometimes in patients with severe diabetic ketoacidosis.[2-4] Until recently, the mechanism was not clear. According to Artis et al.,[110] the growth of *Rhizopus oryzae* was accelerated in serum collected from the patients with severe diabetic ketoacidosis. They explained that the serum pH of the patients was lower due to the disease, and as a result, UIBC is reduced by the low pH.

In accord with this point of view, Abe et al.[111] carried out murine experiments to clarify the relationship between zygomycosis and diabetic ketoacidosis, and succeeded in verifying it in vivo.

Last of all, we refer to the differences in susceptibility to fungal infections among mouse strains. There was a definite difference in susceptibility to *C. neoformans*[112] infection among mouse strains. Thus, we have to choose animals carefully when doing animal experiments.

Animal models of individual mycoses are described by the other authors in the following chapters.

REFERENCES

1. Szaniszlo, P. J., Jacobs, C. W., and Geis, P. A., Dimorphism: morphological and biochemical aspects, in *Fungi Pathogenic for Humans and Animals, Part A*, Biology, Howard, D. H., Ed., Marcel Dekker, New York, 1983, 323.
2. Conant, N. F., Smith, D. T., Baker, R. D., and Callaway, J. L., *Manual of Clinical Mycology*, 3rd ed., W.B. Saunders, Philadelphia, 1971.
3. Emmons, C. W., Binford, C. H., Utz, J. P., and Kwon-Chung, K. J., *Medical Mycology*, 3rd ed., Lea & Febiger, Philadelphia, 1977.
4. Chandler, F. W., Kaplan, W., and Ajello, L., *A Colour Atlas and Textbook of the Histopathology of Mycotic Diseases*, Welfe Medical Publications, London, 1980.
5. Miyaji, M. and Nishimura, K., Studies on arthrospore of *Trichophyton rubrum*, I, *Jpn. J. Med. Mycol.*, 12, 18, 1971.
6. Howard, D. H., Observation on tissue cultures of mouse peritoneal exudates inoculated with *Histoplasma capsulatum*, *J. Bacteriol.*, 78, 69, 1959.
7. Howard, D. H. and Herndon, R. L., Tissue cultures of mouse peritoneal exudates inoculated with *Blastomyces dermatitidis*, *J. Bacteriol.*, 80, 522, 1960.

8. Howard, D. H., Dimorphism of *Sporotrichum schenckii*, *J. Bacteriol.*, 81, 464, 1961.

9. Converse, J. L., Growth of spherules of *Coccidioides immitis* in a chemically defined liquid medium, *Proc. Soc. Exp. Biol. Med.*, 90, 709, 1955.

10. Pine, L. and Webster, R. E., Conversion in strains of *Histoplasma capsulatum*, *J. Bacteriol.*, 83, 149, 1962.

11. Miyaji, M. and Nishimura, K., Parasitic forms of pathogenic fungi, Nihon Rinsho, *Jpn. J. Clin. Med.*, 38, 14, 1980.

12. Miyaji, M., Nishimura, K., and Kuroda, F., Studies on the parasitic forms of *Rhinocladiella pedrosoi. I*, *Jpn. J. Med. Mycol.*, 14, 20, 1973.

13. Miyaji, M. and Nishimura, K., Investigation on dimorphism of *Blastomyces dermatitidis* by agar-implantation method, *Mycopathologia*, 60, 73, 1977.

14. Nishimura, K. and Miyaji, M., Dimorphism of *Microascus cirrosus*, *Mycopathologia*, 90, 29, 1985.

15. Miyaji, M. and Nishimura, K., Conversion of *Coccidioides immitis* from a mycelial form to spherules using the 'agar-implantation method', *Mycopathologia*, 90, 121, 1985.

16. Nishiyama, C., Miyaji, M., Saheki, M., and Morioka, S., Studies on the parasitic forms of *Trichophyton rubrum* isolated from patients with granuloma trichophyticum using the "agar-implantation method", *Jpn. J. Dermatol.*, 12, 325, 1985.

17. Anon., Classification of etiologic agents on the basis of hazard, 5th ed., Center for Disease Control, U.S. Department of Health, Education, and Welfare, Washington, D.C., 1977.

18. Baker, E. E., Mrak, E. M., and Smith, C. E., The morphology, taxonomy, and distribution of *Coccidioides immitis* Rixford and Gilchrist 1896, *Farlowia*, 1, 199, 1943.

19. Tarbet, J. E., Wright, E. T., and Newcomer, V. D., Experimental coccidioidal granuloma. Developmental stages of sporangia in mice, *Am. J. Pathol.*, 28, 901, 1952.

20. Miyaji, M., Nishimura, K., and Ajello, L., Scanning electron microscope studies on the parasitic cycle of *Coccidioides immitis*, *Mycopathologia*, 89, 51, 1985.

21. Sun, S. H., Sekhon, S. S., and Huppert, M., Electron microscopic studies on saprobic and parasitic forms of *Coccidioides immitis*, *Sabouraudia*, 17, 265, 1979.

22. Segretain, G., *Penicillium marneffei* n. sp., agent d'une mycose du systeme réticuloendothélial, *Mycopathol. Mycol. Appl.*, 11, 327, 1959.

23. Garrison, R. G. and Boyd, K. S., Role of the conidium in dimorphism of *Blastomyces dermatitidis*, *Mycopathologia*, 64, 29, 1978.

24. Salazar, M. E. and Restrepo, A., Morphogenesis of the mycelium to yeast transformation in *Paracoccidioides brasiliensis*, *Sabouraudia*, 22, 7, 1985.

25. Maeda, M., Freeze-fracture electron microscopic studies of plasma membrane-changes in *Sporothrix schenckii* under various conditions of cultivation, *Acta Sch. Med. Univ. Gifu*, 30, 927, 1982.

26. Nishimura, K. and Miyaji, M., Studies on the parasitic forms of *Sporothrix schenckii* in scales and crusts. III. Clinical study of thirteen cases, *Jpn. J. Med. Mycol.*, 16, 57, 1975.

27. Moore, M., *Histoplasma capsulatum*: its cultivation on the chorioallantoic membrane of the developing chick and resulting lesions, *Am. J. Trop. Med.*, 21, 627, 1941.

28. Dowding, E. S., The spores of *Histoplasma*, *Can. J. Research*, Section E., 26, 265, 1948.

29. Goos, R. D., Germination of the macroconidium of *Histoplasma capsulatum*, *Mycologia*, 65, 662, 1964.

30. Garrison, R. G. and Boyd, K. S., Electron microscopy of yeast-like cell development from the microconidium of *Histoplasma capsulatum*, *J. Bacteriol.*, 133, 345, 1978.

31. Procknow, J. J., Page, M. I., and Loosli, C. G., Early pathogenesis of experimental histoplasmosis, *A.M.A. Arch. Pathol.*, 69, 413, 1969.

32. Brandt, F. A., Early tissue reactions to a South African strain of *Histoplasma capsulatum* in laboratory animals, *J. Patho. Bacteriol.*, 62, 259, 1950.

33. Miyaji, M. and Nishimura, K., Defensive role of granuloma against fungal infection, in *Filamentous Microorganisms*, Biological Aspects, Arai, T., Ed., Japan Scientific Society Press, Tokyo, 1985, 263.

34. Wortis, H. H., Nehlsen, S., and Owen, J. J., Abnormal development of the thymus in 'nude' mice, *J. Exp. Med.*, 134, 681, 1971.

35. Nieuwerkerk, H. T. M., Lowenberg, B., and van Bekkum, D. W., *In vivo* studies on immunological reconstitution of nude mice, in *Proc. 1st Int. Workshop on Nude Mice*, Rygaard, J. and Povlsen, C. O., Eds., Gustav Fisher Verlag, Stuttgart, 1974, 175.

36. Burns, W. H., Bullups, L. C., and Notkins, A. L., Thymus dependence of viral antigens, *Nature (London)*, 256, 654, 1975.

37. Pelletier, M. and Montplaisir, S., The nude mouse: model of deficient T-cell function, *Methods Achiev. Exp. Pathol.*, 7, 149, 1975.

38. Cauley, L. K. and Murphy, J. W., Response of congenitally athymic (nude) and phenotypically normal mice to *Cryptococcus neoformans* infection, *Infect. Immun.*, 23, 644, 1979.

39. Graybill, J. R. and Drutz, D. J., Host defense in cryptococcosis. II. Cryptococcosis in the nude mouse, *Cell. Immunol.*, 40, 263, 1978.
40. Nishimura, K. and Miyaji, M., Histopathological studies on experimental cryptococcosis in nude mice, *Mycopathologia*, 68, 145, 1979.
41. Miyaji, M. and Nishimura, K., Studies on organ specificity in experimental murine cryptococcosis, *Mycopathologia*, 76, 145, 1981.
42. Watabe, T., Miyaji, M., and Nishimura, K., Studies on relationship between cysts and granulomas in murine cryptococcosis, *Mycopathologia*, 86, 113, 1984.
43. Grosse, G., Mishra, S. K., and Staib, F., Selective involvement of the brain in experimental murine cryptococcosis. II. Histopathological observation, *Bakt. Hyg. I. Abt. Orig. A*, 233, 106, 1975.
44. Shiraishi, A., Nakagaki, K., and Arai, T., Experimental sporotrichosis in congenitally athymic (nude) mice, *J. Reticuloendothel. Soc.*, 26, 333, 1979.
45. Hachisuka, H., Studies on experimental sporotrichosis, *Jpn. J. Dermatol.*, 89, 1053, 1979.
46. Miyaji, M. and Nishimura, K., Defensive role of granuloma against *Sporothrix schenckii* infection, *Mycopathologia*, 80, 117, 1982.
47. Dickerson, C. L., Taylor, R. L., and Drutz, D. J., Susceptibility of congenitally athymic (nude) mice to sporotrichosis, *Infect. Immun.*, 40, 417, 1983.
48. Kariya, H. and Iwatsu, T., Statistical survey of 100 cases of sporotrichosis, *Jpn. J. Dermatol.*, 6, 211, 1979.
49. Miyaji, M., Chandler, F. W., and Ajello, L., Experimental histoplasmosis capsulati in athymic nude mice, *Mycopathologia*, 75, 139, 1981.
50. Edwards, P. Q., and Palmer, C. E., Nationwide histoplasmin sensitivity and histoplasma infection, *Public Health Rep.*, 78, 241, 1963.
51. Newberry, W. M., Jr., Chandler, J. W., Jr., Chin, T. D. Y., and Kirkpatrick, C. H., Immunology of the mycoses. I. Depressed lymphocyte transformation in chronic histoplasmosis, *J. Immun.*, 100, 436, 1968.
52. Kirkpatrick, C. H., Chandler, J. W., Jr., Smith, T. K., and Newberry, W. M., Jr., Cellular immunologic studies in histoplasmosis, in *Histoplasmosis*, Balows, A., Ed., Charles C Thomas, Springfield, Ill., 1971, 371.
53. Smith, J. W., Progressive disseminated histoplasmosis, *Ann. Intern. Med.*, 76, 557, 1971.
54. Graybill, J. R. and Alford, R. H., Variability of sequential studies of lymphocyte blastogenesis in normal adults, *Clin. Exp. Immunol.*, 25, 28, 1976.
55. Williams, D. W., Graybill, J. R., and Drutz, D. J., *Histoplasma capsulatum* infection in nude mice, *Infect. Immun.*, 21, 973, 1978.
56. Vanbreuseghem, R., Étude clinique, mycologique et histopathologique de l'histoplasmose africaine, *Bruxelles-Medical*, 56, 85, 1976.
57. Nishimura, K. and Miyaji, M., Studies on a saprophyte of *Exophiala dermatitidis* isolated from a humidifier, *Mycopathologia*, 77, 173, 1982.
58. Nishimura, K. and Miyaji, M., Defense mechanisms of mice against *Exophiala dermatitidis* infection, *Mycopathologia*, 81, 9, 1983.
59. Nishimura, K. and Miyaji, M., Pathogenicity of *Exophiala jeanselmei* for ddY mice, *Mycopathologia*, 91, 29, 1985.
60. Miyaji, M. and Nishimura, K., Conidial ontogenesis of pathogenic black yeasts and their pathogenicity for mice, *Proc. Indian Acad. Sci. (Plant Sci.)*, 94, 437, 1985.
61. Nishimura, K. and Miyaji, M., Defense mechanisms of mice against *Fonsecaea pedrosoi* infection, *Mycopathologia*, 76, 155, 1981.
62. Kurita, N., Cell-mediated immune responses in mice infected with *Fonsecaea pedrosoi*, *Mycopathologia*, 68, 9, 1979.
63. Iwatsu, T., Miyaji, M., Taguchi, H., Okamoto, S., and Kurita, N., Skin test-active substance prepared from culture filtrate of *Fonsecaea pedrosoi*, *Mycopathologia*, 67, 101, 1979.
64. Cutler, J. E., Acute systemic candidiasis in normal and congenitally thymic-deficient (nude) mice, *J. Reticuloendothel. Soc.*, 19, 121, 1976.
65. Rogers, T., Balish, J. E., and Manning, D. D., The role of thymus-dependent cell-mediated immunity in resistance to experimental disseminated candidiasis, *J. Reticuloendothel. Soc.*, 20, 291, 1976.
66. Binford, C. H., Thompson, R. K., and Gorham, M. E., Mycotic brain abscess due to *Cladosporium trichoides*, a new species, *Am. J. Clin. Pathol.*, 22, 535, 1952.
67. Felger, C. E. and Friedman, L., Experimental cerebral chromoblastomycosis, *J. Infect. Dis.*, 111, 1, 1962.
68. Aravysky, R. A. and Aronson, V. B., Comparative histopathology of chromomycosis and cladosporiosis in the experiment, *Mycopathol. Mycol. Appl.*, 36, 322, 1968.
69. Amma, S. M., Paniker, C. K. J., Iype, P. T., and Rangaswamy, S., Phaeohyphomycosis caused by *Cladosporium bantianum* in Kerala (India), *Sabouraudia*, 17, 419, 1979.

70. Dixon, D. M., Shadomy, H. J., and Shadomy, S., Dematiaceous fungal pathogens isolated from nature, *Mycopathologia*, 70, 153, 1980.
71. Kwon-Chung, K. J. and de Vries, G. A., Comparative study of an isolate resembling Banti's fungus with *Cladosporium trichoides*, *Sabouraudia*, 21, 59, 1983.
72. Nishimura, K. and Miyaji, M., Tissue responses against *Cladosporium trichoides* and its parasitic forms in congenitally athymic nude mice and their heterozygous littermates, *Mycopathologia*, 90, 21, 1985.
73. Miyaji, M. and Nishimura, K., Granuloma formation and killing functions of granuloma in congenitally athymic nude mice infected with *Blastomyces dermititidis* and *Paracoccidiodes dermatitidis*, *Mycopathologia*, 82, 129, 1983.
74. Tabeta, H., Mikami, Y., and Arai, T., Problems in experimental *Candida* infection of congenitally athymic (nude) mice, *Jpn. J. Med. Mycol.*, 21, 256, 1980.
75. Lee, K. W. and Balish, E., Systemic candidosis germ free, flora-defined and conventional nude and thymus-bearing mice, *J. Reticuloendothel. Soc.*, 29, 71, 1981.
76. Lee, K. W. and Balish, E., Systemic candidosis in silica-treated athymic and euthymic mice, *Infect. Immun.*, 41, 902, 1983.
77. Miyaji, M. and Nishimura, K., Murine coccidioidomycosis — reproductive cycle of spherule and tissue responses, *Jpn. J. Med. Mycol.*, 23, 287, 1982.
78. Beaman, L., Pappagianis, D., and Benjamini, E., Significance of T cells in resistance to experimental murine coccidioidomycosis, *Infect. Immun.*, 23, 681, 1977.
79. Clemons, K., Leathers, C. R., and Lee, K. W., Systemic *Coccidioides immitis* infection in nude and beige mice, *Infect. Immun.*, 47, 814, 1985.
80. Nishimura, K. and Miyaji, M., Studies on the growth of *Aspergillus fumigatus* in the brain of mouse, *Jpn. J. Med. Mycol.*, 12, 24, 1971.
81. Miyaji, M. and Nishimura, K., Relationship between proteolytic activity of *Aspergillus fumigatus* and the fungus' invasiveness of mouse brain, *Mycopathologia*, 62, 161, 1977.
82. Kuroda, F., Miyaji, M., Fujiwara, K., and Nishimura, K., Studies on relationship between pathogenicity for mice and proteolytic activity of *Aspergillus fumigatus*, *Jpn. J. Med. Mycol.*, 15, 138, 1974.
83. Arai, T., Shiraishi, A., Yokoyama, K., and Nakagaki, K., Resistance and immune response of nude mice to experimental fungal infections, in *Proc. 3rd Int. Workshop on Nude Mice*, Vol. 1, Oncology, Reed, N. D., Ed., Gustav Fisher, New York, 1979, 77.
84. Corbel, M. J. and Eades, S. M., Experimental mucormycosis in congenitally athymic (nude) mice, *Mycopathologia*, 62, 117, 1977.
85. Warnock, D. W., Immunological and other defects predisposing to fungal infection in the compromised host, in *Fungal Infection in the Compromised Patients*, Warnock, D. W. and Richardson, M. D., Eds., John Wiley & Sons, New York, 1982, 29.
86. Miale, J. B. and Rywlin, A. M., Spleen, in *Pathology*, Vol. 2, 7th ed., Anderson, W. A. D. and Kissane, J. M., Eds., C.V. Mosby, St. Louis, 1977, 1489.
87. King, H. and Shumacker, H. B., Splenic studies: susceptibility to infection after splenectomy performed in infancy, *Ann. Surg.*, 136, 239, 1952.
88. Haller, J. A. and Jones, E. L., Effect of splenectomy on immunity and resistance to major infections in early childhood, *Ann. Surg.*, 163, 902, 1966.
89. Eraklings, A. J., Kevy, S. V., and Diamond, L. K., Hazard of overwhelming infection after splenectomy in childhood, *N. Engl. J. Med.*, 276, 1225, 1967.
90. Shinefield, H. R., Steinberg, C. R., and Kay, D., Effect of splenectomy on the susceptibility of mice inoculated with *Diplococcus pneumonia*, *J. Exp. Med.*, 123, 777, 1966.
91. Whitacker, A. N., The effect of previous splenectomy on the course of pneumococcal bacteriaemia in mice, *J. Pathol. Bacteriol.*, 95, 357, 1968.
92. Leung, L.-S. E., Szal, G. J., and Drackman, R. G., Increased susceptibility of splenectomized rats to infection with *Diplococcus pneumoniae*, *J. Infect. Dis.*, 126, 507, 1972.
93. Coil, J. A., Dickerman, J. D., and Boulton, E., Increased susceptibility of splenectomized mice to infection after exposure to an aerosolized suspension of type III *Streptococcus pneumonia*, *Infect. Immun.*, 21, 412, 1978.
94. Goldthorn, J. F. and Schwartz, A. D., Poor protective effect of unregenerated splenic tissue to pneumococcal challenge after splenectomy, *Surg. Forum*, 29, 469, 1978.
95. Ghadirian, E. and Meerovitich, E., Effect of splenectomy on the site of amoebic liver abscesses and metastatic foci in hamsters, *Infect. Immun.*, 31, 571, 1981.
96. Llende, M., Santiago-Delpin, E. A., and Lavergne, J. A., Effect of splenectomy on *Trypanosoma lewisi* infection on young rats, *Infect. Immun.*, 40, 1127, 1983.
97. Skamene, E. and Chayasirisobhon, W., Enhanced resistance to *Listeria monocytogenes* in splenectomized mice, *Immunology*, 33, 851, 1977.
98. Eling, W. M. C., Role of spleen in morbidity and mortality of *Plasmodium berghei* infection in mice, *Infect. Immun.*, 30, 635, 1980.

99. Robinett, J. P. and Rank, R. G., Splenomegaly in murine tripanosomiasis: T cell-dependent phenomenon, *Infect. Immun.*, 23, 270, 1979.
100. Miyaji, M. and Nishimura, K., Increased resistance of splenectomized mice to *Sporothrix schenckii* infection, *Mycopathologia*, 96, 143, 1986.
101. Abe, F., Fujioka, Y., Nakamura, N., and Ommura, Y., A pathological study on deep fungus infections in autopsy cases with leukemia and lymphoreticular disease, *Jpn. J. Med. Mycol.*, 22, 185, 1981.
102. Uetsuka, A., Satoh, S., and Ohno, Y., Protective effect of PSK, A protein-bound polysaccharide preparation against candidiasis in tumor-bearing mice, *Adv. Exp. Med. Biol.*, 121B, 21, 1979.
103. Mardon, D. N. and Robinnette, E. H., Jr., Organ distribution and viability of *Candida albicans* in noncancerous and tumor-bearing (Lewis lung carcinoma) mice, *Can. J. Microbiol.*, 24, 1515, 1978.
104. Johnson, J. A., Lau, B. H. S., Nuater, R. L., Slatar, J. M., and Winter, C. E., Effect of L1210 leukemia on the susceptibility of mice to *Candida albicans* infections, *Infect. Immun.*, 19, 146, 1978.
105. Abe, F., Tateyama, M., Tada, M., Shibuya, H., and Ommura, Y., Leukemia and mycosis. I. Experimental candidiasis in leukemic mice, *Jpn. J. Med. Mycol.*, 24, 133, 1983.
106. Abe, F., Tateyama, M., Shibuya, H., and Ommura, Y., Leukemia and mycosis. II. Experimental aspergillosis in leukemic mice, *Jpn. J. Med. Mycol.*, 25, 325, 1984.
107. Weinberg, E. D., Iron and infection, *Microbiol. Rev.*, 42, 45, 1978.
108. Abe, F., Tateyama, M., Shibuya, H., Azumi, N., and Ommura, Y., Experimental candidiasis in iron overload, *Mycopathologia*, 89, 59, 1985.
109. Abe, F., Tateyama, M., Shibuya, H., and Ommura, Y., Experimental aspergillosis in iron overload, *Jpn. J. Med. Mycol.*, 25, 290, 1984.
110. Artis, W. M., Fountain, J. A., Delcher, H. K., and Johnes, H. F., A mechanism of susceptibility to mucormycosis in diabetic ketoacidosis. Transferrin and iron availability, *Diabetes*, 31, 1109, 1982.
111. Abe, F., Shibuya, H., Tateyama, M., and Ommura, Y., Experimental mucormycosis associated with diabetic ketoacidosis — role of iron and serum cofactors, in *Abstr. 9th Int. Congr. ISHAM*, Atlanta, Ga., May, 1985.
112. Rhodes, J. C., Wicker, L. S., and Urba, W. J., Genetic control of susceptibility of *Cryptococcus neoformans* in mice, *Infect. Immun.*, 29, 494, 1980.
113. Miyaji, M. and Nishimura, K., unpublished data.

Chapter 2

ANIMAL MODELS OF CUTANEOUS AND SUBCUTANEOUS MYCOSES

S. Watanabe

TABLE OF CONTENTS

I. CHROMOMYCOSIS AND PHAEOHYPHOMYCOSIS

A. Introduction

Chromomycosis is a chronic granulomatous disease caused by dematiaceous fungi, and is characterized clinically by verrucous skin lesions with brown, spherical, thick-walled muriform cells (sclerotic cells). Dissemination from the skin lesions to regional lymph nodes or to internal organs via the blood stream occurs in rare instances. Though the term chromoblastomycosis is used in the texts by Conant et al.[1] and Rippon,[2] and chromomycosis is employed in the book by Emmons et al.,[3] the terms are usually used synonymously. Ajello,[4] proposed that chromoblastomycosis, exhibiting sclerotic cells as the parasitic forms, should be distinguished from phaeohyphomycosis, showing hyphal elements in subcutaneous mycoses caused by dematiaceous fungi, and that the term chromomycosis should be eliminated.

The fungi causing classical chromomycosis include *Fonsecaea pedrosoi, F. compactum, Phialophora verrucosa, Cladosporium carrionii,* and *Rhinocladiella aquaspersa.* Subcutaneous abscesses or phaeomycotic cysts are formed by *Exophiala jeanselmei, E. moniliae, E. spinifera, P. richardsiae, P. parasitica, P. repens,* and *Drechslera spicifera.*[5] *Wangiella (Exophiala) dermatitidis,* frequently reported in Japan, usually produces warty plaques with sclerotic cells in tissues of the skin and mucosa, but a rare cystic lesion in a finger due to this agent has been reported.[6] In this case also, fungal elements appeared as sclerotic cells. However, Matsumoto et al. insisted that these organisms in tissues of reported cases due to *W. dermatitidis* are not typical sclerotic cells.[7] On culture media, this fungus forms two typical colonies of granular to yeast-like form and mycerial form, and is neurotropic. Nishimura and Miyaji regarded the term *"Exophiala"*, advocated by de Hoog, to be more suitable than *"Wangiella".*[21] *Cladosporium bantianum (trichoides)* specifically attacks the brain,[8] but may cause secondary lesions in the skin.[9]

Dematiaceous fungi inducing chromomycosis, such as *P. verrucosa, F. pedrosoi, E. jeanselmei,* and *W. dermatitidis,* have been isolated from soil and humidifiers;[10,11] *C. bantianum* also grows in soil or on bark, and has been isolated from soil in Japan.[12]

Table 1
FUNGUS ELEMENTS IN TISSUE IN EXPERIMENTAL ANIMALS[5]

	Rat, mouse intraperitoneally	Mouse intravenously		
Lesion	Peritoneal cavity	Brain	Lung	Kidney
Organisms				
F. pedrosoi	SC Mycel.	Mycel.	Mycel. SC (few)	Mycel.
P. verrucosa	Mycel.			
W. dermatitidis				
Granular form	SC	SC	SC	—
Mycelial form	Spore shape SC (few)	Mycel.	Mycel. SC (few)	Mycel. SC (few)
C. bantianum	Mycel.	Mycel.	Mycel.	Mycel.

Note: Mycel., Mycelial Form and SC, sclerotic cells.

From Fukushiro, R., *Handbook of Dermatology*, Nakayama Shoten, Tokyo, 1982, 110.

The fungi is introduced generally by trauma into the skin, but the primary lesion may occur in the lung through the inhalation of spores. Especially, *C. bantianum* is considered to gain entry through the lung.

According to Fukushiro,[5] in 24 of 318 dematiaceous fungal infection cases reported in Japan, the patients died of lesions that developed only in internal organs, or from dissemination from the skin and mucosa to internal organs. Especially, 14 of 15 children died, indicating the extremely poor prognosis of this infection in childhood. In those who died of invasion of internal organs, the causative agents were *F. pedrosoi* (nine cases), *W. dermatitidis* (nine cases), *P. verrucosa* (one case), and *C. bantianum* (two cases). The mortality rate in these patients was 9/241 (3.7%) for *F. pedrosoi* infection; 9/23 (39.1%) for *W. dermatitidis* infection; 1/6 (16.7%) for *P. verrucosa* infection, and 2/2 (100%) for *C. bantianum* infection. Though the high mortality rate for *C. bantianum* infection is well-known, the rate for *W. dermatitidis* infection, occurring frequently in Japan, is also high.

Natural infection of chromomycosis among animals has been reported in dogs, cats, horses, and frogs.[2] Animals susceptible to experimental chromomycosis are mice, rats, rabbits, hamsters, guinea pigs, and monkeys, the first two having the highest susceptibility.[13] The fungi often used in animal experiments include *F. pedrosoi* and *W. dermatitidis*, which are frequently isolated from humans and *C. bantianum* causing a high mortality. Fungal elements in experimental animals studied by Fukushiro[5] are shown in Table 1.

B. Animal Models of Chromomycosis and Phaeohyphomycosis
1. Fonsecaea pedrosoi Infection
Polak[14] induced a chronic infection in cortisone-pretreated mice by intravenous (i.v.) inoculation of *F. pedrosoi*. The manifestations were characterized by black lesions in the skin and subcutis with sclerotic cells resembling the tissue form of human chromomycosis in addition to lesions in the brain and other organs. The procedure was as follows. *F. pedrosoi* was incubated for one week on Sabouraud's dextrose agar at 30°C, the surface growth was harvested, ground in a mortar, filtered through a fine cotton mesh to remove debris, and the filtrate was adjusted for an optical density (8 ×

10^7 colony-forming units per 0.2 mℓ). Male Swiss-albino mice were injected with 0.2 mℓ of this suspension into a lateral tail vein. The mice were treated with 2 mg cortisone subcutaneously 90 min before and 24 hr after inoculation of the fungus. During the first 3 to 4 weeks after inoculation, no symptoms of illness were detected. Thereafter, the legs became swollen and black spots appeared on the tail. These spots increased in size, and after 7 weeks, the tail and the legs became dark and numerous nodules appeared. KOH preparations from subcutaneous abscesses in animals revealed abundant hyphal growth and the typical roundish, thick-walled, sclerotic cells. The lung, brain, and kidney also showed granulomatous lesions and microabscesses, whereas the liver and spleen in most animals were free of lesions. In the brain, the fungus grew only in its hyphal form.

Borelli[15] used new-born litters (about 7 days old). Inocula were prepared by grinding 3-week-old colonies of *F. pedrosoi,* and suspending them in isotonic saline solution. The amount of the suspension injected intraperitoneally was adjusted to 1/10th to 1/15th of the animal's body weight. During the first month after the inoculation, the pathogens were detected mainly in thoracic and abdominal organs, and occasionally in the brain. Thereafter, dark nodules were observed along the tail, at the tibio-tarsal joint. The hind limbs subsequently became immobile, followed by cachexia and death. *F. pedrosoi* was present in the form of septate cells.

Kurita[16] intravenously injected 5-week-old male ddy mice with yeast-like cells of *F. pedrosoi* strain Tsuchiya, and studied the time course of the cellular and humoral immune responses. Viable fungus cells were recovered from several organs from the 14th to 16th day, and from the brain on the 36th day after inoculation. Inflammatory lesions were observed in the brain, lung, heart, liver, spleen, kidney, and intestine during the first 30 days after inoculation. Nishimura and Miyaji[17] studied the defense mechanisms of athymic nude (nu/nu) mice and their heterozygous (nu/+) littermates of BALB/c background against *F. pedrosoi* infection following inoculation into a tail vein of 0.1 mℓ of 0.02, 0.1, and 0.5% fungus cell suspension. The survival rate of 10 nu/nu mice, inoculated with the 0.5% cell suspension, was 0% by the 16th day after inoculation, and that of nu/+ mice, 100% by the 25th day.

Male and female white rats also were used in some experiments.[5,18,19] A suspension (0.05 mℓ) adjusted to a hemacytometer count of 2×10^7 conidial cells per milliliter of *F. pedrosoi* was injected subcutaneously into the dorsum of the left foot of a rat. Using these models, skin tests for chromoblastmycosis or histopathology studies were performed.

2. Wangiella dermatitidis Infection

Polak[14] produced an acute, fetal cerebral infection closely resembling human phaeohyphomycosis in mice. The yeast form cells of *W. dermatitidis* were suspended in 0.15 *M* saline and counted in a hemocytometer. Male Swiss-albino mice were injected i.v. with 6×10^6 colony-forming units. Six hours after inoculation, viable fungal cells were found in all tissues examined. During the subsequent 4 days, the viable counts decreased significantly in the lung and increased significantly in the kidney and remarkably in the brain. Most of the animals died within the first week, due to the rapid multiplication of fungus in the brain. About 20% of the infected animals recovered during the 2nd or 3rd week. In these animals, significantly lower cell counts in the brain were observed.

Nishimura and Miyaji[21] studied the defense mechanisms of mice against *E. dermatitidis* infection using congenitally athymic nude (nu/nu) mice and their heterozygous (nu/+) littermates of BALB/c background. Each mouse was inoculated intravenously with 10^6 yeast cells of the fungus. Viable cells were counted in the brain, kidney, and liver of the nu/+ and nu/nu mice. Furthermore, various internal organs were examined histopathologically. The susceptibility of the nu/nu mice to the fungus was higher than that of the nu/+ mice. Viable cells in the brain of the nu/+ mice reached a peak on the 4th day, and thereafter decreased gradually. In the nu/nu mice, they reached a peak on the 4th day, remained at this level until the 25th day, and thereafter increased gradually.

Jotisankasa et al.[22] investigated the virulence of this fungus using normal and predisposed mice injected intraperitoneally. All strains studied were neurotropic. The pathogenicity of this fungus in laboratory animals has been demonstrated by other investigators.[20,23,24] Iwatsu et al.[19] found that all male rats infected subcutaneously with *W. dermatitidis* exhibited positive reactions to the same fungus antigen.

3. Cladosporium bantianum Infection

In our experiment, *C. bantianum* isolated from the first reported clinical case in Japan was used.[8] Intravenous injection of 2 to 3×10^5 fungus cells (conidia and hyphal fragments), obtained from culture, caused the death of white mice in 1 to 2 weeks. The fungus was detected in both smears and cultures of the brain lesions. According to Fukushiro,[5] mice and rats are susceptible to this fungus, and in the brain and other organs, only the mycelial form was observed. (Table 1)

4. Other Species Infection

P. verrucosa and *E. jeanselmei* infected rats were used for evaluation of skin test using antigens prepared from culture filtrates of the same fungus studied by Iwatsu et al.[19] Intravenous inoculation of *P. verrucosa* produced systemic infection involving most internal organs.[2] Fukushiro et al. found only mycelial elements of this fungus in tissue of *P. verrucosa* infected mice and rats, and no sclerotic cells.[5]

C. carrionii is not neurotropic, and this characteristic is used to distinguish it from *C. bantianum* in animal experiments.[2]

II. DERMATOPHYTOSES

A. Introduction

Dermatophytosis is the most common fungal infection, and the lesions are usually superficial, involving keratinized layers of the skin, hair, and nails.[25] When dermatophytes affect the dermis of humans as well as animals, severe inflammation, as observed in kerion celci, occurs and abscesses are formed in the inflammatory reaction to eliminate the fungi. In immunosuppressed individuals, the fungi may induce the formation of granulomatous tumors in subcutaneous tissues, or, in extremely rare cases, invade various internal organs, leading to death of the host.[26] Therefore, unlike common animal models of other fungal infections, it is important to produce local skin-infection models of dermatophytosis. Such models have been produced usually in guinea pigs by various techniques.[27-29] Though these models are useful for studying the pathogenicity and the virulence of fungi, and especially for observing their parasitic state in the hair (ectothrix or endothrix), some species may be pathogenic only for humans and cannot be successfully inoculated in animals.

Animal models are employed also for investigation of pathophysiology and immune mechanisms in which they are subjected to various examinations such as the trichophytin skin test. More importantly, excellent animal models are needed for the development of new antifungal agents, because drugs exhibiting potent anti-fungal actions in vitro must undergo in vivo trials before clinical application.

B. Animal Models of Dermatophytoses

1. Animals

Small animals such as guinea pigs, rabbits, mice, rats, and hamsters, and more infrequently, large animals such as cats, dogs, fowl, monkeys, cattle, and horses are used for inoculation experiments. Guinea pigs are most commonly employed because they are readily infected with dermatophytes, showing strong inflammatory reactions, and the course of infection is relatively uniform regardless of their sex. However, these models are not always satisfactory since the course of infection is too short to be compared with that of human chronic infections. Therefore, the course of skin lesions should be prolonged using highly virulent strains of *T. mentagrophytes*.[29] According to some investigators,[30,31] the duration of lesions of dermatophytosis in guinea pigs can be prolonged by using corticosteroids, whereas Fisher and Sher[32] observed no difference in the severity or duration of infection between guinea pigs treated with steroids and those untreated. Iwata[29] reported that various attempts to reduce the resistance of the host, such as administration of steroids or immunosuppressants, had little effect. Green et al.[33] grafted guinea pig skins into congenitally athymic (nude) mice, inoculated them with *T. mentagrophytes* after 30 days, and demonstrated chronic dermatophyte infection persisting for 58 days.

Rabbits are also suitable for immunoserological studies of dermatophytosis due to the relative ease of blood collection.[34] However, the lesions are less inflammatory and the course of infection is more variable than in guinea pigs. Reiss[35] described that *Trichophyton purpureum (T. rubrum)* infection of the skin could be maintained longer in castrated rabbits than in control animals.

Studies of experimental dermatophytosis in dogs have been few. Kushida[36] observed the initial occurrence of lesions in white mongrel dogs 4 to 6 days after inoculation. The latent period was comparable to that in guinea pigs, but the duration between the fastigium and crust exfoliation was considerably longer (15 to 30 days). Patches of alopecia with scales persisted for 1 to 3 months, and spontaneous healing required 3 to 4 months or longer. The course of infection was prolonged by the inoculation, with not only *T. mentagrophytes* but also *M. canis* and *M. gypseum,* and unaffected by the inoculation method such as sandpaper method, one-point prick inoculation, and topical application of the inoculum with a knife (Figure 1). These results are consistent with the report by Georg et al.,[37] who observed dogs inoculated with *T. mentagrophytes* for 30 to 50 days.

2. Organisms

Among many species of dermatophytes (Table 2), zoophilic dermatophytes, especially *T. mentagrophytes* var. *mentagrophytes (T. asteroides, T. granulare), T. quinckeanum,* and *M. canis* have been frequently used since they are highly pathogenic for experimental animals and produce marked lesions. Some of the geophilic dermatophytes such as *M. gypseum* have similar degrees of pathogenicity against animals. *T. quinckeanum* characteristically produces scutula on the skin of mice and guinea pigs, but sometimes fails to do so. This fungus, which has been frequently used for animal experiments,[38] is classified as a member of the *T. mentagrophytes* complex because of the recent success in mating with *Arthroderma benhamiae,* one of teleomorphs of *T. mentagrophytes.*[39] According to Plempel and Bartman[40] superficial dermatophytosis produced in guinea pigs by *T. quinckeanum* infection resembles human ringworm, and can be utilized for evaluation of the efficacy of topical preparations. They classified the infection lesions into four stages based on the occurrence of erythema and scutula during the course. Chittasobhon and Smith[41] inoculated guinea pigs with *T. mentagrophytes* var. *erinacei* isolated from a hedgehog and observed symptoms and a course

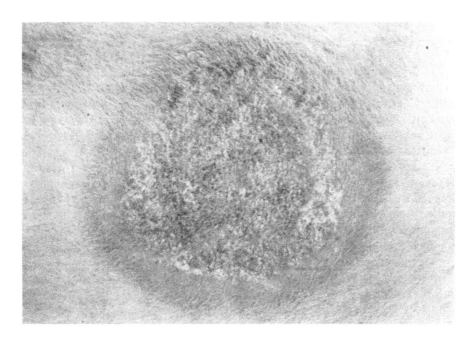

FIGURE 1. Experimental dermatophytosis in a dog, 20th day after inoculation with *Microsporum gypseum.* (From Kushida, T., *Bull Nippon Vet. Zootech. Coll.*, 24, 64, 1975. With permission.)

Table 2
EXPERIMENTAL DERMATOPHYTOSIS

	References by animal					
Organism	Guinea pig	Mouse	Nude mouse	Rabbit	Cat	Dog
Zoophilic Dermatophytes						
T. mentagrophytes	29—32,34,36		33	34	37	36,37
var. mentagrophytes	40,41,45—47,					
var. granulare	48,49,50,51,52,53,					
(*T. asteroides*)	56,57,59—62					
var. erinacei	41					
var. quinckeanum	40,53,54	38				
T. verrucosum	56,57					
M. canis	30,41,44,51					36
M. persicolor	41					
Geophilic Dermatophytes						
M. gypseum	41,51			34		36
Anthropophilic Dermatophytes						
T. rubum	35,41,42	55		35		
(*T. purpureum*)						
T. tonsurans	41					
M. audouinii	44					
E. floccosum	41					

similar to those seen in experimental ringworm by *T. mentagrophytes* var. *mentagrophytes.*

Anthropophilic dermatophytes are generally less capable of producing infection in animals. The rate of positivity after inoculation with *T. mentagrophytes* var. *interdigitale,* an important fungus responsible for tinea pedis, is low, and even when positive, inflammation is extremely mild, healing spontaneously within 20 days. *T. rubum* is the fungus most frequently isolated from human ringworm but has been considered to have a relatively low ability to induce experimental ringworm in guinea pigs. Reiss[35] produced *T. rubrum* infection with a prolonged course in castrated rabbits. Takahashi[42] inoculated 47 strains of *T. rubrum* into the guinea pig back and 10 of them induced infection. The entire course of infection was about 3 to 4 weeks and the progression of infection resembled that caused by *T. mentagrophytes.* All ten strains produced exothrix infection with three also showing endothrix infection.

Smith[43] found no difference in virulence between parasitically formed arthrospores and saprophytic aleuriospores or conidia. On the other hand, Klingman et al.[44] observed higher positive rates by inoculating animals with arthrospores of *M. canis* and *M. audouini* infected hair than with conidia of the fungi collected from culture.

When stock strains are used for experiments, those with strong virulence must be selected by a passage in animals.

3. Inoculation

The inoculation routes with dermatophytes can be classified generally as epidermal or extraepidermal. Epidermal inoculation to animals is done usually in the skin of the back, but sometimes in the extremities, auricula, tail, or cock's comb. Extraepidermal inoculation is performed by subcutaneous, intramuscular, intraperitoneal, intravenous, or intracardiac injection.

a. Epidermal Inoculation of Dermatophytes
i. Abrasion Method (Sandpaper Method)

Hartley-strain white guinea pigs (350 to 600 g) are the most suitable, though rabbits and dogs can also be used; non-white guinea pigs, however, are inappropriate for observation. Some circular hairless patches (3 to 5 cm in diameter) are produced in the back by plucking hair by hand. According to Takahashi,[28] clipping of hair with scissors or shaving with a razor permits early regrowth of hair, making observation of the lesion difficult. After abrading the sites lightly enough with sterilized sandpaper so as not to cause bleeding, a fungal suspension is applied. Inoculation can also be made by abrading the skin with a lancet or rubbing a spore suspension into the depilated skin (about 5 cm²) with a scalpel to achieve both abrasion and inoculation simultaneously.[45]

For the preparation of spore suspensions, a powdery portion of well-grown colonies of a fungus is harvested from culture in 1/10 Sabouraud's agar with salt (Takashio's medium),[39] cultured at 27°C for 2 to 4 weeks, and agitated while adding physiological saline. The spores are adjusted to predetermined concentrations using a Toma's counting chamber. Under optimal conditions, as few as 10 spores of *T. mentagrophytes* can produce lesions, but 10^4 or more conidia are usually needed.[43] In experiments for evaluation of the efficacy of anti-fungal agents, high concentrations (10^7 to 10^8 per mℓ) are commonly used to ensure the development of lesions.

When highly pathogenic *T. mentagrophytes* is inoculated, erythema and scales occur at the site of inoculation after 3 to 6 days, followed by crust formation. The fastigium of the infection is reached after 2 weeks, and persists for 3 to 4 days. Inflammation gradually reduces thereafter, and heals spontaneously within 4 to 6 weeks. Macroscopic observation of the lesion developing at the inoculation site must be performed

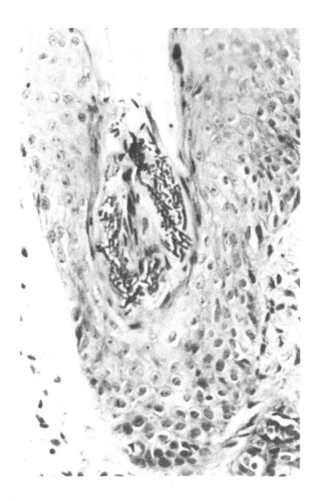

FIGURE 2. Experimental dermatophytosis in guinea pig; hair follicle and hair invaded with hyphae, 14th day after epidermal inoculation of *Trichophyton mentagrophytes*. (Periodic acid, Schiff stain.)

together with direct microscopy of KOH-treated skin specimens or histological examination to determine whether the lesion is caused by the inoculated organisms. The presence of the fungus in hair follicles and hair serves as a definite evidence of the infection (Figures 2 and 3). Furthermore, the growth of the same strain as that inoculated must also be confirmed by the isolates from the lesion.

ii. Gum Tape Method (Cellophane Tape Method)

The gum tape method is often used as a modification of the sandpaper method. Hair of guinea pigs is trimmed short with an electric hair clipper, and gum tape[29] or cellophane tape[28] cut out in circles (2 to 3 cm in diameter) is applied to a few sites of the skin and pulled off. This procedure is repeated until the horny layer of the epidermis is nearly detached. The inoculum, adjusted to predetermined concentrations, is dropped on the center of the sites with an injector (0.05 m*l* for each site). The procedure is completed as the suspension spreads over the site. The course of the infection is similar to that by the sandpaper method.

FIGURE 3.   A lesion on the back of a guinea pig, 9th day after inoculation with *T. mentagrophytes* *(A. benhamiae* "−" Strain) using the sandpaper method.

iii. One-Point Prick Inoculation

Fungal mats are collected from a colony and divided into small pieces using needles. One point of the depilated skin of the guinea pig is pricked with a needle smeared with a small amount of the inoculum. A lesion develops 4 to 7 days after inoculation with *T. mentagrophytes.* The erythemato-squamous lesion enlarges in a circular shape (Figure 4), reaching a maximum size in about 2 weeks, and heals spontaneously after about 1 month. Takahashi and Tanuma[46] described that intracutaneous trichophytin reaction became positive after 7 to 10 days, reaching a maximum level after 12 to 14 days, and ringworm allergy persisted about 50 days (22 to more than 170 days). These authors also found that the courses in guinea pigs after reinoculation with *T. mentagrophytes* were different from those after the first inoculation, and classified the former into two types. In one type, fungal elements were demonstrated in the lesions produced by reinoculation. The lesions developed after 3 to 5 days and healed after 11 to 21 days. This type of reaction, indicating an abortive form of the initial infection, was seen in a few animals. The other type, found in the majority, was an aseptic type; no fungal elements were demonstrated in the lesions, and punctate erythema developed after 2 to 4 days and healed after 9 to 21 days.

Since the ringworm lesions produced by this method are not influenced by inflammation due to trauma as those produced by the abrasion method, the enlarging process of the lesions per se can be serially observed in details and numerically expressed. This method is, therefore, suitable for assessment of the effects of anti-fungal agents.[47] Furthermore, it is technically simple and requires no dressing of the inoculation site, but the rate of positive inoculation by this method is slightly lower than that by the sandpaper method. However, virulence of a fungus may be evaluated conveniently by utilizing the difference in the rate of positive inoculation between these two methods. Watanabe and Hironaga[48] demonstrated that virulence of *Arthroderma vanbreusegh-emii,* a telemorph of *T. mentagrophytes,* differs in guinea pigs among its three forms (granulosum-asteroides, powdery, and persicolor) associated with the differences in

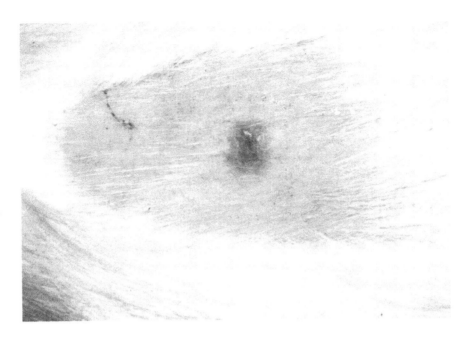

FIGURE 4. A scaly erythematous lesion on the back of a guinea pig, 9th day after inoculation with *T. mentagrophytes (A. benhamiae* "−") using the one-point prick method.

their sexual power. No differences were detected in pathogenicity between "+" and "−" mating types of the granulosum-asteroides form of *A. vanbreuseghemii.*

iv. Occlusive Dressing Method

For induction of experimental infection, the surface of the skin should be kept adequately moist with occlusive dressing.[49,50] Tagami et al.[51] succeeded in producing a highly reproducible lesion without traumatization of the skin by occluding dermatophyte spores. *T. mentagrophytes (T. asteroides)* was used and the following standardized method was established. The organism is cultured on Sabouraud's dextrose agar at 28°C for 2 weeks. After sterile distilled water was added, a colony was dislodged with a platinum-wire loop. Clumps of fungus in the suspension were grounded in a mortar with a pestle. The suspensions in test tubes were vibrated by an electric mixer for 5 min and filtered through 8 layers of gauze to exclude large fragments. The filtrates were washed with sterile water 3 times by centrifugation at 3000 rpm for 10 min. The preparations thus obtained were composed of numerous small conidia and a few hyphal fragments. For adjustment of concentrations, conidia were counted using a hemocytometer in the same manner as in counting white blood cells. As experimental animals, guinea pigs (Hartley strain) weighing 350 to 500 g were used. An absolute condition of an occlusive state in the back skin is complete depilation of the site. Therefore, after plucking hair, any remaining short hair must be trimmed by an electric clipper. The conidial suspension (0.01 m*l*) is placed over the depilated skin with a micropipette. The site is covered with a sheet of polyethylene film (2 × 2 cm), and the inoculum is evenly spread with the fingers covered with a glove. An elastic bandage (3.8 cm in width) is applied around the trunk; the bandage should be long enough to ensure adhesion of the polyethylene film to the inoculation site and also not to be removed by animals. If the bandage is too tight, animals are debilitated, and if too loose, they soon remove it with hind legs (Figure 5). The occlusive dressing is removed usually after 24 hr. The severity of the lesions increased with prolongation of the occlusion period.

FIGURE 5. Inoculation of a guinea pig using the occlusive dressing method. An elastic bandage is bound around the trunk.

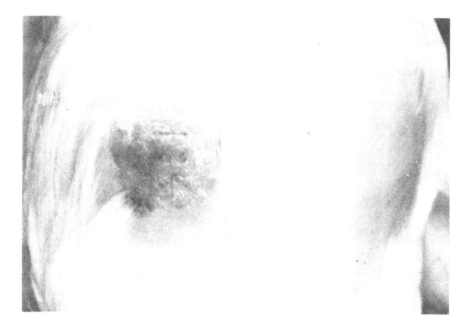

FIGURE 6. *T. mentagrophytes* infection in a guinea pig using the occlusive dressing method, 12th day after the inoculation of 24 hr: (right) with granulosum-asteroides strain; (left) with persicolor-form strain of *A. vanbreuseghemii.* Inoculation is positive only in the right site.

Minor erythema was occasionally seen after a 24-hr occlusion. Papuloerythematous lesions confined to the inoculation site appeared after 2 or 3 days, followed by scales after a few days. The extent of the lesions was evaluated by using a 3 point scale: 1+ represents a few punctate papuloerythematous lesions; 2+, scattered erythematous patches in addition to 1+ lesions, and 3+, entirely diffuse erythematous lesions over the site of inoculation. Even in 3+ lesions, the maximum peripheral enlargement was twice the size of the inoculation area. The most intense reaction was seen after 6 to 12 days, when swelling and spreading erythematous lesions increased rapidly with thickened scales and occasional bloodcrust (Figure 6). The intracutaneous trichophytin reaction and patch test with trichophytin, which were negative before the inoculation, became positive during this period. Inflammation in the lesions began to resolve thereafter and healed after 4 weeks, leaving alopecic scars.

When animals positive for trichophytin test were reinoculated, as was reported by Bloch,[52] lesions developed on the 1st or 2nd day, and inflammation reached the climax on the 4 to 8th day. Even at the climax, the inflammation was much milder than that after the first infection, and subsided after 10 days.

Five groups of guinea pigs were inoculated with different doses of *T. mentagrophytes* spores (80, 400, 2000, 10,000, and 50,000 spores/cm² skin area). Each group consisted of 7 animals. Only 1 of the 7 animals was infected with a dose of 80 spores/cm², but the rate of infection increased with dosage. A dose of 10,000 spores/cm² could induce 3+ lesions in all animals. This finding shows that reproducible experimental ringworm can be produced by occlusion of 10,000 spores/cm² for 24 hr. Relative pathogenicity among species: similar experiments with *M. gypseum* and *M. canis* showed that these fungi have clearly lower pathogenicity than *T. mentagrophytes.*

b. Extracutaneous Inoculation of Dermatophytes

Saeves[29] inoculated guinea pigs intravenously with *T. mentagrophytes (T. gypseum)* and *T. mentagrophytes (T. quinckeanum)* and succeeded in producing widespread cutaneous lesions. Bloch[52] described skin lesions in guinea pigs after intracardiac injection of dermatophytes, which healed spontaneously within 3 weeks. Sulzberger[54] reported the recovery of *T. mentagrophytes (T. quinckeanum)* from the heart, liver, kidney, spleen, and lung of guinea pigs after intracardial injection. Sternberg et al.[55] inoculated mice intraperitoneally with *T. rubum* and found chronic granulomatous lesions in the omentum, liver, spleen, and muscles. Recently, Van Cutsem et al.[56,57] described generalized dermatophytosis that affected all parts of the skin and internal organs in experimental animals following systemic administration of dermatophytes. They used *T. mentagrophytes* var. *granulare,* strain B-32663, for i.v. route infection. Inoculum concentrations ranging from 117 to 60,000 colony-forming units (CFU) per gram body weight were injected in a lateral vena of the penis of male albino guinea pigs. Ringworm lesions developed in the animals. Macroscopically visible fungal eruptions appeared after 2 to 4 days, depending on the inoculum size. The climax was reached in 2 weeks and remained stable for 6 to 7 weeks after injection. The animals remained highly infected until the end of the 11th week. At a dose of 117 CFU/g body weight, a few minor dermatophyte lesions occurred without involvement of internal organs. At 15,000 CFU/g body weight, all animals presented skin dermatophytosis and 91% of the lung cultures were positive. Fungal granulomas were formed in the lungs and persisted for the entire experimental period. Using these models, Van Cutsem[58] tested new antifungal substances.

C. In Vivo Evaluation of the Effects of Antifungal Agents

Animal models of dermatophytoses are of great value for the development of new antifungal agents. Various evaluation methods employing the above-mentioned experimental models of dermatophytoses have been used by many investigators. Representative methods are described below.

1. "Kaken" Method

This "Kaken" method reported by Sakai et al.[59] has been widely used in Japan. The back skin of guinea pigs is depilated and abraded at 4 to 6 sites to produce patches of skin injuries of the size of a coin. After application of a suspension of *T. mentagrophytes,* a test agent is applied once a day from the 2nd, 3rd, and 4th day for 6, 7, or 8 consecutive days, respectively. These tests are termed the 2nd-, 3rd-, and 4th-day methods. The symptoms are recorded daily. On the first day, after completion of the application of the test agent, guinea pigs are sacrificed and washed well with warm water and soap. Three patches (5 mm in diameter) are collected from each lesion with sterilized scissors. Each patch is divided into halves, placed on Sabouraud's dextrose agar, and cultured at 26°C for 1 week. The rate of negative culture is regarded as the cured rate, by which the efficacy of the agent is evaluated. Some sites were left untreated, with the solvent alone, for controls. If an agent with known efficacy is used as the control, the virulence of the strain can be also assessed. Though this method is suitable for evaluation of anti-ringworm agents with marked fungicidal effects, it may be inappropriate for fungistatic drugs. Some of the antifungal agents effective for the treatment of human ringworms may be quite ineffective in fungicidal tests in cultures. Culturing tests are suited for evaluation of synthetic chemicals with potent fungicidal effects, but can be too rigorous for antibiotics with fungistatic actions. The pharmaceutical effects of antifungal agents are considered to be more appropriately evaluated in terms of the rate of effectiveness based on the measurement of lesions produced by inoculation.

2. Modified "Kaken" Method

In the "Kaken" method, treatment is initiated before manifestation of symptoms. Therefore, the prophylactic, rather than therapeutic, effects of the agent are evaluated by this method. Recently, however, a modified method in which the treatment is initiated after confirmation of the lesion has been widely practiced. Tawara et al.[60] applied a suspension of higly virulent *T. mentagrophytes* SM-110 (*Arthroderma vanbreuseghemii* "+") to the back skin of guinea pigs by the sandpaper method, and recorded lesion scores for 40 days by observing the natural course after inoculation. According to the descriptions by Weinstein et al.[61] and Gordee and Matthews,[62] the severity of lesions was graded 0 to +4: 0 represents no lesions; +1, a stage in which only a few small erythematous papulae are seen or lesions are in the process of healing with new hair growth; +2, a stage in which scattered erythematous patches or slight growth of new hair is observed despite peripheral reddening; +3, a stage in which diffuse erythema is seen over the entire site of inoculation with abundant scales or thick crusts; and +4, a stage in which inflammation is at the climax and is accompanied by bleeding. Test agents are applied once a day (0.5 g) for 10 consecutive days from the 5th day of infection. Drug efficacy was determined according to the final score of the lesion on the day of completion of the treatment and the negative rate in the culture test. The culture test was performed according to the "Kaken" method. Guinea pigs were sacrificed on the day after termination of the treatment. The skin of the drug-treated and untreated control patches was excised, washed in warm water to remove excess drug, and cut into five tissue fragments. The tissues were cultured on Sabouraud's dextrose agar supplemented with cycloheximide (500 μg/mℓ) and penicillin G (100 μg/mℓ). The negative rate was expressed as the percentage of skin samples negative for the fungus after culture at 28°C for 10 days.

Final lesion scores — Comparison between the drug-treated and untreated control sites was made using Freeman-Halton's direct probability calculation method.

Rates of negative cultures — Differences between the treated and untreated control sites as well as those among drugs were evaluated using Freeman-Halton's direct-probability calculation method or analysis based on empirical logistic conversion (Cox's method; Bonferroni's universal error rate).

3. Evaluation of the Effects of Antifungal Agents Using One-Prick Inoculation Method

Tanuma[47] evaluated drug efficacy based on the degree of inhibition of the enlargement of experimental ringworm lesions. The size of erythema on the 11th day after inoculation with *T. asteroides* was 10 to 13 mm in diameter (mean, 10.9 mm) in the controls, but it was less than 8 mm when drugs with known efficacy were applied once a day for 8 consecutive days from the 3rd day. Therefore, a drug is considered to be effective if the size of erythema was less than 8 mm in diameter. This method appears to be useful to evaluating fungistatic effects of drugs since the results of this evaluation are consistent with clinical efficacy of the drugs which show a cure rate of zero by the Kaken method.

4. Oral Treatment of Superficial Skin Mycosis and Generalized Mycosis in Guinea Pigs

Epidermal inoculation models in guinea pigs are useful for in vivo evaluation of not only topical skin treatment but also oral administration of antifungal substances such as ketoconazole and griseofulvin and intravenous treatment with agents such as Miconazole. Van Cutsem[58] described oral treatment of skin mycosis with various agents starting at day −1 (prophylactically) or day +3 (therapeutically) for 14 consecutive days daily. In disseminated dermatophytosis induced by intravenous administration of *T. mentagrophytes* (15,000 CFU/g body weight), small inflammatory lesions appeared on the skin after 3 to 4 days and developed into specific ringworm lesions within one week.[56,57] Ketoconazole (2.5 to 20 mg/kg) was orally administered daily for 21 to 30 consecutive days from day 0. Animals infected with dermatophytes were sacrificed 12 to 21 days after the last treatment and cultures of various organs were made.[58]

III. MYCETOMA

A. Introduction

Mycetoma is a clinical entity characterized by the development of tumefactions, multiple sinuses, and the formation of granules. Mycetomas are classified into two major groups by the causative organisms; actinomycotic mycetoma (actinomycetoma) due to actinomycetes and eumycotic mycetoma (eumycetoma) due to true fungi. The term, Maduromycosis or madura foot is used as a synonym of mycetoma[63,64] or only for eumycotic mycetoma.[65,66]

Mycetoma is prevalent in the tropical and subtropical regions such as in Africa, Central and South America, and India, and often occurs after trauma. The disease is also called mycetoma pedis since the feet are the predominant site of lesions, but the hands, thighs, knees, trunk, and scalp are also occasionally involved. Males are more affected than females, and the incidence is high in those working outdoors. One or several painless subcutaneous nodules develop and the lesion gradually extends deeper sometimes even into bone, followed by formation of multiple sinuses that drain thin pus. The pus contains small granules varying in color according to the species. In actinomycotic mycetoma, white-yellow granules are produced by *Nocardia brasilliensis*, *N. asteroides*, *N. caviae*, and *Actinomadura madurae;* yellow-brown granules by *Streptomyces somaliensis;* red by *A. pelletierii;* and black by *S. paraguayensis.* In eumycotic mycetoma, the granules of *Pseudallescheria boydii*, *Neotestudina rosatii*, *Acremonium spp.,* and *dermatophytes* are white-yellow and those of *Madurella mycetomatis*, *M. grisea*, *Pyrenochaeta romeroi*, *Leptosphaeria senegalensis*, and *Exophiala jeanselmei* are brown-black. Granules in eumycotic mycetoma can be readily differentiated from those in actinomycetic mycetoma since they are generally larger and consist of thicker hyphae. Regardless of the fungal agents, the lesions are pathohistologically similar. One to several granules are occasionally surrounded by localized cellular-infiltration lesions in the dermis consisting of lymphocytes, macrophages, and a large accumulation of neutrophils, around which epithelioid and plasma cells are observed.

B. Experimental Mycetoma in Laboratory Animals

1. Actinomycetoma

Some works have indicated differences in the degree of pathogenicity for three *Nocardia* agents of mycetoma in experimental infection in laboratory animals: *N. brasiliensis* is the most virulent and always produces granules; *N. caviae* is less virulent but still produces granules, and *N. asteroides* is significantly less pathogenic and produces no granules.[64] Zlotnik et al.[67] also reported similar results in *N. asteroides*-infected mice. In Japan, *N. asteroides* is a common causative agent of mycetoma, though the number of patients is small.[68] Fukushiro et al.[69] succeeded in producing infection in animals by inoculation with *N. asteroides* and *N. caviae.* In rabbits inoculated with *N. asteroides*, granules with marked clubs in the margin were observed frequently in pulmonary lesions but not (or rarely) in other organs. Especially in the brain and kidney, only mycelial growth was seen. The bacterial elements in tissue did not differ between animals and humans. In guinea pigs, rats, and mice, which were positive after inoculation with *N. asteroides* as well as *N. caviae*, the bacterial elements were observed as all granules, frequently accompanied by clubs. Destombes et al.[70] also reported granule formation often associated with clubs in guinea pigs inoculated with *N. asteroides.* The above-mentioned findings are inconsistent with the reports by Emmons et al.[63] and Rippon[64] showing that *N. asteroides* does not produce granules, and also cast doubt on the commonly accepted view that granules in nocardiosis are not accompanied by clubs. Thus, regional differences may exist in the pathogenicity of isolated strains.

a. Nocardia brasiliensis Infection

Conde et al. injected into the footpad of mice 0.05 m*l* of 0.15 *M* NaCl containing 2 to 5×10^7 viable *N. brasiliensis* organism. After day 14, postinfection, mycetoma were found. The lesions were located in the dermis and subcutaneous tissue, and contained multiloci having well-defined zones, each with a central mass of organisms forming a granule of 50 to 100 μm in diameter.[71,72] The footpads of mice were also injected with 0.1 m*l* of a suspension containing viable *N. brasiliensis* in incomplete Freund's adjuvant (IFA) and, depending on the concentration of organisms, mycetoma with granules was observed after 15 to 20 days or 1 to 3 months.[67,73]

b. Nocardia caviae Infection

Beaman et al.[74] injected single-cell suspensions of *N. caviae* into normal, athymic, and asplenic mice by several different routes: intravenously, intraperitoneally or subcutaneously in the footpad, and intranasally. Chronic, progressive disease leading to the formation of mycetomas was found only in mice injected intravenously. Kurup and Sandhu[50] examined the pathogenicity of *N. caviae* isolated from soil and demonstrated that intraperitoneal inoculation with smaller concentrations (1:320 or 1:640, v/v) into white mice produced multiple abscesses on the abdominal wall and viscera. Rabbits and guinea pigs receiving intraperitoneal or intravenous inoculation of organisms also developed abscesses. Local abscess formation was found in guinea pigs receiving subcutaneous inoculation. The granules were cream-colored, spherical to oval, soft, and 200 to 500 μm in diameter. Similar results were reported by Fukushiro et al.[69] using rabbits, rats, mice, and guinea pigs and by Zlotnik and Buckley[67] using BALB/c mice.

c. Nocardia asteroides Infection

N. asteroides has been frequently reported to produce lesions but not granules in animal experiments.[63,64,67] However, Fukushiro et al.[69] demonstrated granules, with or without clubs, after inoculating animals with five strains of *N. asteroides* isolated from man. The inoculum was prepared either by adding physiological saline (10 mℓ) to fungal colonies obtained by culture on Sabouraud's dextrose agar slants for one month or adding saline (30 mℓ) to a fungal suspension (5 mℓ) obtained by a shaking culture on Sabouraud's dextrose broth at 37°C for 8 days. The suspensions were injected intravenously into rabbits (1 to 2 mℓ), or intraperitoneally into rats (0.2 to 0.4 mℓ), mice (0.1 mℓ), and guinea pigs (0.5 mℓ). All animals registered positive after the inoculation. The highest positive rate and the most marked granule formation were observed in rats. The animals were killed after 1 month, and numerous, small intra-abdominal nodules were seen. In rabbits, many granules were present in pulmonary lesions. Reiss[77] found that guinea pigs and rabbits are susceptible to *Nocardia asteroides* infection; in that study, a well-tritiated suspension (3 to 7 mℓ) was injected either intraperitoneally into guinea pigs or intravenously (into an ear vein) into rabbits. Intravenous injection was repeated at weekly intervals for 3 to 5 weeks. Two intraperitoneal injections given within 1 week were generally sufficient to produce the disease in guinea pigs. Some animals developed extensive visceral lesions with granules. Destombes et al.[70] also observed granule formation in guinea pigs inoculated with *N. asteroides.*

d. Actinomadura madurae Infection

Rippon et al.[76] produced experimental mycetoma caused by *A. madurae* in cortisone treated mice and untreated mice. In mice, intraperitoneally given cortisone acetate (1 mg), *A. madurae* microcolonies (50 mg) were injected into the right groin area. By the 10th day, large subcutaneous swellings occurred at the injection sites. Histological examination showed a mixed pyogenic and granulomatous infection and the formation of granules (100 to 500 μm). Untreated mice were inoculated with 1×10^6 viable organisms, and the disease course was similar to that in cortisone-treated mice.

2. Eumycetoma
a. Madurella mycetomatis and Other Species Infection

Though some species of true fungi have been reported to be eumycotic agents of mycetoma in humans and animals, reports on successful inoculation of experimental animals are rare. Even with *Madurella mycetomatis*, a fungus most frequently isolated from human mycetoma, successful inoculations are few;[78,80] in addition, reproducibility by the reported methods has been questionable.[79] Symmers[82] subcutaneously inoculated fresh granules obtained from a human hand with mycetoma caused by *Exophiala jeanselmei* into rabbits and after 58 days excised one subcutaneous nodule, in which granules similar to those in the tissue of the patient were observed. Avram[80,81] intraperitoneally inoculated mice with a suspension of *Acremonium kiliense (Cephalosporium falciforme)* isolated from a patient with maduromycosis and found granule formation in 18% of the 50 mice. Infection associated with granules was also observed after inoculation with *Pseudallescheria boydii* and *M. mycetomatis.* Mycetoma caused by *Aspergillus* and dermatophyte, which has attracted less attention, is described below.

b. Aspergillus Infection

Jidoi and Matsumoto[83] reported a 50-year-old male with mycetoma caused by *A. fumigatus.* Subcutaneous hard nodules and sinuses were seen in the buttocks after repeated injections of oily penicillin. The granules were observed in the abcesses located in the deep area, and *Aspergillus fumigatus* was demonstrated by culture. Though mycetoma induced by this fungus is extremely rare, it has been produced in animal experiments. Fukushiro et al.[84] intraperitoneally inoculated mice with a suspension of *A. fumigatus* (0.2 mℓ) isolated from a patient with primary pyoderma-like aspergillosis and observed typical granules in one nodule.

Table 3
EXPERIMENTAL SPOROTRICHOSIS

	Reference numbers
Animals	
Mice	88—90, 96, 99—101, 104—106, 113
Nude mice	107, 108
Rats	93, 102—104, 106, 109, 112
Rabbits	103, 110, 112
Guinea pigs	96, 110, 112
Hamsters	90, 97
Cats	98
Dogs	110, 111
Monkeys	102, 106
Inoculation	
Intraperitoneal	88, 89, 93, 96, 100—105, 109
Intravenous	103, 107, 108, 112
Intracardial	93, 102
Intratesticular	102, 106, 109
Intramuscular	
(Hind thigh)	105
Subcutaneous	
(Footpad or	89, 97, 102, 103
Tail)	113
Epidermal or	96, 106
intradermal	

c. Dermatophyte Infections

Burgoon et al.[85] reported a human case of mycetoma due to *T. rubrum* with granules on the dorsum of the right foot. Fukushiro et al.[66] intraperitoneally inoculated mice with *T. rubrum* isolated from a patient with granuloma trichophyticum and observed granule formation in a small nodule. They also reported a 20-year-old male with mycetoma due to *M. ferrugineum* in the buttocks discharging granules.[66] Scalp mycetoma caused by the same fungus has been found in cases from Africa.[86] In Australia, a 9-year-old boy with scalp mycetoma showing white granules was described, but whether isolated *M. canis* and *T. mentagrophytes* were causative agents is unclear.[87]

IV. SPOROTRICHOSIS

A. Introduction

Sporotrichosis (Table 3) is a chronic fungal infection caused by *Sporothrix schenckii*. *S. schenckii* grows in nature and has been cultured directly from soil and plants. However, some investigators have demonstrated that a few saprophytic isolates in nature are pathogenic for mice.[88,89] Mariat reported that strains of *Ceratocystis stenoceras*, usually parasitic on plants, increased in virulence by animal passage, and was pathogenic for the hamster and mouse. This led him to consider that the perfect stage of *S. schenckii* might be *C. stenoceras*.[90]

S. schenckii is a dimorphic fungus. On Sabouraud's dextrose agar at room temperature, it forms a conidia-producing hyphal fungus in the mycelial phase, and on brain-heart infusion (BHI) agar at 37°C, it is converted to the yeast phase, In humans, the fungus is generally introduced into the skin by traumatic implantation. Human sporotrichosis occurs mostly from September to April in Japan and from April to July in the southern hemisphere when ambient temperature is low.[91,92] Mackinnon et al. reported that sporotrichosis developed in the bones of the paws and tail only in animals raised under a low ambient temperature of 5 to 20°C, and no lesions were seen in these sites in animals kept at 31°C.[93]

The clinical types in humans are divided into lymphocutaneous, fixed cutaneous, extracutaneous, and disseminated sporotrichosis.[94] In the lymphocutaneous type, the initial lesion appears as a granulomatous nodule, subsequently multiple subcutaneous nodules along the course of the local lymphatics, then followed by the development of ulcerations. In fixed cutaneous sporotrichosis, the lesion is restricted to the site of inoculation without lymphangitic spread. These two cutaneous types are common, but extracutaneous and disseminated sporotrichosis, suggesting blood stream infection, are extremely rare.[95] Kwon-Chung reported that conidia from eight isolates obtained from fixed cutaneous lesions (nonlymphangitic) formed colonies at 35°C but failed to do so at 37°C, although conidia from 26 clinical isolates obtained from lymphocutaneous or extracutaneous lesions formed colonies at both 35°C and 37°C in vitro within 5 days. She also showed that 37°C-type isolates, intraperitoneally injected into mice, grew in the internal organs but multiplied faster in the testes when the animals were kept at 25°C, while the 35°C-type isolates failed to multiply in the internal organs but grew well in testes.[96]

Natural infection of sporotrichosis among animals has been documented in horses, dogs, cats, boars, rats, mules, foxes, camels, dolphins, and fowl.[94] Experimental infections with *S. schenckii* have been reported in rats, mice, rabbits, guinea pigs, dogs, cats, hamsters, and monkeys, the first two being most suitable as experimental models. However, most of these infections were induced by intraperitoneal inoculations that initiated a systemic disease, a manifestation rarely seen in humans.[95] Charoenvit and Taylor[97] described a hamster model for self-limited, lymphatic infection resembling the classical disease in humans, and a model for a systemic disease which disseminated from an initial subcutaneous infection. Barbee et al. showed that adult cats can be readily infected and that the course of the disease in many aspects resembles that of human sporotrichosis.[98]

B. Animal Models of Sporotrichosis
1. Mice Models
a. Introduction

Animal models of *Sporothrix schenckii* infections are commonly produced by intraperitoneal inoculation of mice or rats for evaluation of the virulence of the fungus. Such models are useful particularly for comparative study of virulence between isolates from humans or animals and those from soil or plant.[88,89,100] These models, in which the course of a progressive and systemic disease can be serially observed, are also employed for assessing the therapeutic value of antifungal agents.[99]

b. Materials and Methods

Male white mice weighing about 20 g are usually employed, but female mice have also been used.[99,100] Some investigators recommend that the animals should be housed at 20°C since the rate of infection is higher at lower temperatures.[89,93]

The test strains are cultured at 26 to 28°C for 1 to 2 weeks on Sabouraud's dextrose agar plates and harvested in the mycelial growth phase. Suspensions of the mycelial-phase are made by dispersing the cells in a sterile mortar while adding sterile physiologic saline, and this is then filtered through thick, sterile gauze. Saline suspensions containing 10^6 conidia in 0.2 to 0.5 ml are inoculated intraperitoneally into mice. These conidial suspensions may be contaminated by mycelial components. Suspensions are usually given without supplementation, but in a few studies, they were mixed with an equal volume of sterile and neutral 5% aqueous gastric mucin to promote pathogenesis.[88,100] Organisms in the yeast phase grown on BHI agar at 37°C may be more satisfactory, but most strains isolated from nature do not convert to the yeast phase. The mycelial-phase cells, therefore, have been used in a number of studies.[88,89,96,100]

For cells in the yeast phase, culturing is done on BHI dextrose agar slants at 37°C for 7 days, and then the cells are suspended in sterile phosphate-buffered saline solution. After being washed two times in the solution by centrifugation, they are resuspended, and adjusted to predetermined concentrations by using a hemacytometer.[107] The suspension, containing 10^5 to 10^6 yeast cells in 0.2 mℓ, is inoculated intravenously into mice.

Pus or exudate from a lesion obtained from a patient may be used directly without culturing.[106] Identification of this fungus in the human pus is difficult by direct examination under the microscope. However, extensive sporotrichosis can be induced in mice by injection of the pus diluted with saline.

Mice are sacrificed at various intervals and each organ is examined by culture and histopathological techniques. If the inoculated fungus is pathogenic, severe peritonitis or orchitis is observed within 2 to 3 weeks followed by dissemination to the mesenteric lymph nodes, spleen, liver, kidney, lung, and heart. Nodulation in the tail is also a common finding. The histopathological feature in early stages is abscess formation, which becomes granulomatous about 1 week after the infection. Kinbara and Fukushiro[104] histopathologically studied the nodules that developed on the surface of testicles, as well as in the greater omentum, and noted numerous fungal elements in microabscesses within these nodules. The spores varied in size, and were spherical or oval, and sometimes budding. The cigar-shaped cells were scanty, contrary to earlier reports. Asteroid bodies were also found in testicular lesions.

c. Inoculation Route

The fungal cells may be introduced by routes other than intraperitoneal injection. Strains that were negative after inoculation by one route may be positive after inoculation by another route.

Subcutaneous injection of a cell suspension (0.2 mℓ) in the back of the mouse hind-leg induces swelling after about 2 weeks, becoming prominent after 3 weeks.[89] For intramuscular inoculation, a cell suspension (0.25 mℓ) is injected in the medial side of the right hind-thigh of white mice. A visible swelling of the infected thigh is observed within 7 to 12 days after inoculation, and peak swelling is generally observed within 20 to 30 days. On excision, the infected thigh reveals abscess formation in the muscles.[105] Following intratesticular inoculation, the rat testicles were swollen and covered with small yellow nodules within 4 weeks.[102,106,109] When given intravenously, the cell suspension (0.2 mℓ) is injected into the tail vein.[107]

2. Nude Mice Models

Athymic (nude) mice are useful for studying the defense mechanism of mice against *S. schenckii* infection.[108] According to Miyaji and Nishimura, granuloma formation is observed in nude mice (nu/nu) from early stages of infection, but, unlike in heterozygous mice (nu/+), growth of the cells is limited in the granuloma.[107]

3. Rat Model

The male white rat is also suitable as a model for experimental sporotrichosis. Reiss[109] examined the effects of intratesticular or intraperitoneal inoculation of 0.5 to 1.0 mℓ of a heavy spore suspension of the mycelial or yeast cells. He found that within 14 to 20 days, the inoculation is usually followed by abscess formation in the pelvic region and later, of the mesenteric lymph nodes, spleen, and liver, and frequently ulcerative lesions develop along the tail at this stage. The animals rarely survive until the 4th week. Benham,[102] describing the rats that received intratesticular injections, noted that their testicles were swollen and covered with small yellow nodules after 4 weeks.

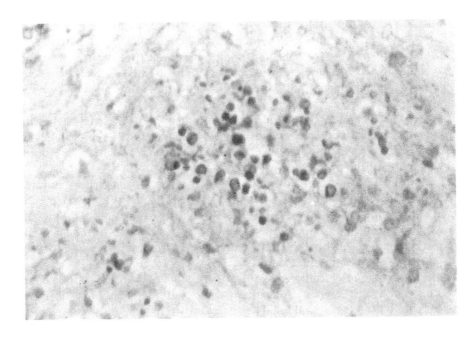

FIGURE 7. Section from experimental orchitis of white rat, 21st day post-inoculation with *Sporothrix schenckii*. Spherical, oval, and cigar-shaped fungal elements are observed. (Periodic acid, Schiff stain.)

Kobayashi[103] inoculated white rats with a cell suspension (0.5 to 2.0 m*l*) intraperitoneally, subcutaneously, or intramuscularly, and presented detailed description of the course of the disease. Results of intratesticular inoculation have been reported by investigators such as Benham[102] and Hopkins.[106] Mackinnon[93] injected a suspension of cells in the yeast phase intraperitoneally (1 m*l*) or intracardially (0.25 m*l*) in rats, and found a relationship between the pathogenesis and the ambient temperature. Kinbara and Fukishiro[104] intraperitoneally inoculated a suspension of mycelial-phase cells in rats, and observed a greater variety of fungal elements, including the asteroid tissue form, than in mice.

4. Other Models

Rabbit and guinea pig — Rabbits and guinea pigs are infrequently employed for inoculation experiments with *S. schenckii* since the pathogenesis and the course of the disease are more variable than in mice or rats.[103,112] Kwon-Chung produced skin lesions in guinea pigs by epidermal or intradermal inoculation of conidia.[96]

Hamster — Mariat[90] studied the virulence of the fungus in the hamster. Charoenvit and Taylor[97] uniformly induced two types of infection by subcutaneous footpad inoculation of *S. schenckii* in male Syrian hamsters weighing 100 g: a self-limited, lymphatic infection resembling the classical disease in humans, and a generalized nonfatal infection. An infecting dose of 5.3×10^3 yeast cells produced the localized subcutaneous lymphatic disease which was limited to a single limb. In contrast, a 1,000-fold increase in the inoculum (5.3×10^6) produced a systemic infection involving the liver and spleen. Using these models, they investigated the serological response in localized and systemic sporotrichosis.

Cat — Barbee et al.[98] showed that adult cats can be readily infected and that the course of disease, in many aspects, resembles that of human sporotrichosis. Domestic cats were inoculated with a yeast cell suspension of *S. schenckii* isolated from humans into one of the rear footpads. Primary wart-like lesions appeared at the inoculation site approximately 5 weeks after injection of the organism, and secondary lesions generally appeared along the course of afferent lymphatic vessels leading from the foot to the popliteal lymph node. Initially, these lesions were nodular but soon softened and became ulcerated.

Monkey — Benham[102] inoculated monkeys with a suspension (0.5 ml) of mycelial-phase organisms subcutaneously into the left forefinger, and observed two nodules at the site of inoculation within 4 weeks. However, no generalized symptoms developed. Hopkins and Benham[106] also produced multiple nodular lesions closely simulating those seen in lymphangitic form of the disease in man.

Dog — Experimental sporotrichosis has been described by Hekton and Perkins[110] and Kren and Schramek.[111]

C. In Vivo Evaluation of the Effects of Antifungal Agents on Sporotrichosis

Treatment with test agents is initiated intraperitoneally, intravenously, or orally on the day after inoculation of animals with a suspension of *S. schenckii*. The effect of the administration is examined by comparing the survival of treated animals with that of the controls administered the solvent alone by the same route. The animals are sacrificed after various intervals to examine the number of colony-forming units in various organs such as the liver and spleen. The effect of administration is also evaluated by the presence or absence of pathogen. For example, subcutaneous inoculation in the tail of untreated animals produces nodular lesions after 3 days, which markedly enlarged after 14 days, but not in the animals in which the treatment was initiated upon inoculation.[113]

REFERENCES

1. Conant, N. F., Smith, D. T., Baker, R. D., and Callaway, J. L., Chromoblastomycosis, in *Manual of Clinical Mycology*, 3rd ed, Conant, N. F., et al., Eds., W. B. Saunders, Philadelphia, 1971, 503.
2. Rippon, J. W., Chromoblastomycosis and related dermal infections caused by dematiaceous fungi, in *Medical Mycology: The Pathogenic Fungi and the Pathogenic Actinomycetes*, 2nd ed., Rippon, J. W., Ed., W. B. Saunders, Philadelphia, 1982, 260.
3. Emmons, C. W., Binford, C. H., Utz, J. P., and Kwon-Chung, K. J., Chromomycosis, in *Medical Mycology*, 3rd ed., Emmons, C. W. et al., Eds., Lea & Febiger, Philadelphia, 1977, 386.
4. Ajello, L., The gamut of human infections caused by dematious fungi, *Jpn. J. Med. Mycol.*, 22, 1, 1981.
5. Fukushiro, R., Chromomycosis, in *Handbook of Dermatology*, 7A Infections Dermatosis II, Yamamura, Y., Kukita, A., Sano, S., Seiji, M., Eds., Nakayama Shoten, Tokyo, 1982, 110.
6. Greer, K. E., Gress, G., Coopor, P. H., and Harding, S. A., Cystic chromomycosis due to *Wangiella dermatitidis*, *Arch. Dermatol.*, 115, 1433, 1979.
7. Matsumoto, T. and Matsuda, T., Current concept of black yeasts infection, *Hifu Rinsho.*, 26, 1175, 1984.
8. Hironaga, M. and Watanabe, S., Cerebral Phaeohyphomycosis caused by *Cladosporium bantianum*: a case in a female who had cutaneous alternatiosis in her childhood, *Sabouraudia*, 18, 229, 1980.
9. Fukumoto, A., Mizuhara, T., and Matsuda, Y., A case of chromomycosis due to *Cladosporium trichoides*, *Jpn. J. Med. Mycol.*, 23, 31, 1980.
10. Iwatsu, T., Miyaji, M., and Okamoto, S., Isolation of *Phialophora verrucosa* and *Fonsecaea pedrosoi* from nature in Japan, *Mycopathologia*, 75, 149, 1981.
11. Nishimura, K. and Miyaji, M., Studies on a saprophyte of *Exphiala dermatitidis* isolated from a humidifier, *Mycopathologia*, 77, 173, 1982.
12. Matsushima, T., *Icones Microfungorum a Matsushima Lectorum*, Kobe, Japan, 1975, 33.
13. Reiss, F., Experimental mycotic infections on laboratory animals, in *Medical Mycology*, Simons, R. D. G., P., Ed., Elsevier, London, 1954, 50.

14. Polak, A., Experimental infection of mice by *Fonsecaea pedrosoi* and *Wangiella Dermatitidis, Sabouraudia,* 22, 167, 1984.
15. Borelli, D., A method for producing chromomoycosis in mice, *Trans. Soc. Trop. Med. Hyg.,* 66, 793, 1972.
16. Kurita, N., Cell-mediated immune responses in mice infected with *Fonsecaea pedrosoi, Mycopathologia,* 68, 9, 1979.
17. Nishimura, K. and Miyaji, M., Defence mechanisms of mice against *Fonsecaea pedrosoi* infection, *Mycopathologia,* 76, 155, 1981.
18. Aravsky, R. A. and Aronson, V. B., Comparative histopathology of chromomycosis and cladosporiosis in the experiment, *Mycopathologia,* 36, 322, 1968.
19. Iwatsu, T., Miyaji, M., Taguchi, H., and Okamoto, S., Evaluation of skin test for chromoblastomycosis using antigens prepared form culture filtrates of *Fonsecaea pedrosoi, Phialophora verrucosa, Wangiella dermatitidis* and *Exophiala jeanselmei, Mycopathologia,* 77, 59, 1982.
20. Kano, K., Über die Chromoblastomykose durch einen noch nicht als Pathogen Beschriebenen Pilz: *Hormiscium dermatitidis* n sp., *Arch. Dermatol. Syph.,* 176, 282, 1937.
21. Nishimura, K. and Miyaji, M., Defense mechanisms of mice against *Exophiala dermatitidis* infection, *Mycopathologia,* 81, 9, 1983.
22. Jotisankasa, V., Neilsen, H. S. Jr., and Conant, N. F., *Phialophora dermatitidis:* its morphology and biology, *Sabouraudia,* 8, 98, 1970.
23. Shimazono, Y., Isaki, K., Torii, H., Otsuka, R., and Fukushiro, R., Brain abscess due to *Hormodendrum dermatitidis* (Kano) Conant, 1953 — Report of a case and review of the literature, *Folia Psychiatr. Neurol. Jpn.,* 17, 80, 1963.
24. Tsai, C. Y., Lu, Y. C., Wang, L. T., Hue, T. L., and Sugn, J. L., Systemic chromoblastomycosis due to *Hormodendrum dermatitidis* (Kano) Conant, *Am. J. Clin. Pathol.,* 46, 103, 1966.
25. Rippon, J. W., Ed., *Medical Mycology: The Pathogenic Fungi and the Pathogenic Actinomycetes,* 2nd ed., W. B. Saunders, Philadelphia, 1982, 154.
26. Hironaga, M., Okazaki, N., Saito, K., and Watanabe, S., *Trichophyton mentagrophytes* granulomas, *Arch. Dermatol.,* 119, 482, 1983.
27. Reiss, F., Experimental mycotic infections on laboratory animals, in *Medical Mycology,* Simons, R. D. G. P., Ed., Elsevier, London, 1954, 50.
28. Takahashi, S., The techniques of experimental trichophytia, *Rinshou Hifu* (in Japanese), 22, 1219, 1968.
29. Iwata, K., Dermatophytosis (guinea pig), in *Handbook of Animal Models for Human Diseases,* Kawamata, J. and Matsushita, H., Eds., Ishiyaku Pub., Tokyo, (in Japanese), 1982, 280.
30. Goss, W. A., Actor, P., Jambor, W. P., and Pagano, J. F., The *Trichophyton mentagrophytes* and *Microsporum canis* infection of the guinea pigs, *J. Invest. Dermatol.,* 40, 299, 1963.
31. Maestrone, G., Sadek, S., and Mitrovic, M., Lesions of dermatophytosis in guinea pigs treated with triamcinolone acetate: an animal model, *Am. J. Vet. Res.,* 34, 833, 1973.
32. Fisher, M. and Sher, A. M., Virulence of *Trichophyton mentagrophytes* infecting steroid-treated guinea pigs, *Mycopathologia,* 47, 1, 1972.
33. Green, F., Lee, K. W., and Balish, E., Chronic *T. mentagrophytes* dermatophytosis of guinea pig skin grafts on nude mice, *J. Invest. Dermatol.,* 79, 125, 1982.
34. Watanabe, S., Immunobiological studies on dermatomycosis (parts 2 and 3), *Arch. Dermat. (Kyoto),* 54, 106, 1959.
35. Reiss, F., Successful inoculation in animals with *Trichophyton purpureum, Arch. Dermatol. Syph.,* 49, 242, 1944.
36. Kushida, T., Studies on dermatophytosis in dogs. 1. Experimental dermatophytosis in dogs, *Bull. Nippon Vet. Zootech. Coll.,* 24, 64, 1975.
37. Georg, L. K., Roberts, C. S., Menges, R. W., and Kaplan, W., *Trichophyton mentagrophytes* infections in dogs and cats, *J. Am. Vet. Med. Assoc.,* 130, 427, 1957.
38. Hay, R. J., Experimental dermatophytosis: the clinical and histopathologic features of a mouse model using *Trichophyton quinckeanum* (Mouse Favus), *J. Invest. Dermatol.,* 81, 270, 1983.
39. Vanbreuseghem, R., De Vroey, C., Takashio, M., *Practical Guide to Medical and Veterinary Mycology,* 2nd ed., Vanbreuseghem, R. et al., Eds., Masson, New York, 1978, 130.
40. Plempel, M. and Bartman, K., Experimentalle Untersuchungen zur Antimykotischen Wirkung von clotrimazole *in vitro* und bei lokaler Applikation *in vivo, Arzneim. Forsch.,* 22, 1280, 1972.
41. Chittasobhon, N. and Smith, J. M. B., The production of experimental dermatophyte lesions in guiea pigs, *J. Invest. Dermatol.,* 73, 198, 1979.
42. Takahashi, S., Morphological, biological and physiological studies of *Trichophyton rubrum.* I. Morphological studies and experimental inoculation of guinea pigs, *Jpn. J. Dermatol.,* 72, 50, 1962.
43. Smith, J. M. B., The pathogenesis of ringworm, in Proc. VIIIth. Congr. Int. Soc. for Human and Animal Mycology, Baxter, M., Ed., Massey University, New Zealand, 1982, 13.

44. Kligman, A. M., The pathogenesis of tinea capitis due to *Microsporum audouini* and *Microsporum canis, J. Invest. Dermat.*, 18, 231, 1952.

45. Matsuda, A., Ishiki, Y., Yonekura, Y., and Iwata, K., Fundamental studies on antifungal activity of Griseofulvin. III. On curative and fungicidal effects of Griseofulvin in vivo and its hemolytic action, *Jpn. J. Med. Mycol.*, 2, 50, 1961.

46. Takahashi, Y. and Tanuma, R., Dermatophytes allergy, *Jpn. J. Allergol.*, 1, 211, 1952.

47. Tanuma, S., Studies on the screening test of antifungal antibiotics using experimental trichophytosis, *Jpn. J. Med. Mycol.*, (in Japanese), 2, 117, 1961.

48. Watanabe, S. and Hironaga, M., Differences or similarities of the clinical lesions produced by "+" and "−" types members of the "*mentagrophytes*" complex in Japan, in *Sexuality and Pathogenicity of Fungi*, Vanbreuseghem, R. and Vroey, C., Eds., Masson, Paris, 1981, 83.

49. Greenberg, J. H., King, R. H., Kerbs, S., and Field, R., A quantitative dermatophyte infection model in the guinea pig — a parallel to the quantitative human infection model, *J. Invest. Dermatol.*, 67, 704, 1976.

50. Kerbs, S. and Allen, A. M., Effect of occlusion on *Trichophyton mentagrophytes* infection in guinea pigs, *J. Invest. Dermatol.*, 71, 301, 1978.

51. Tagami, H., Watanabe, S., and Ofuji, S., *Trichophyton* contact sensitivity in guinea pigs with experimental dermatophytosis induced by a new inoculation method, *J. Invest. Dermatol.*, 61, 237, 1973.

52. Bloch, B., Allgemeine und experimentelle Biologie der durch Hyphomyceten erzeugten Dermatomykosen, in *Handbuch der Haut und Geschlechtskrankheiten, Elfter Band, Dermatomykosen*, Jadassohn, J., Ed., Springer-Verlag, Berlin, 1928, 321.

53. Saeves, I., Experimentelle Beitrage zur Dermatomykosenlehre, *Arch. Dermatol. Syph.*, 121, 161, 1916.

54. Sulzberger, M. B., Experimentelle Untersuchungen über die Dermatotropie der Trichophytonpilze, *Arch. Dermatol. Syph.*, 157, 345, 1929.

55. Sternberg, T. H., Tarbet, J. E., Newcomer, V. D., and Winter, L. H., Deep infection of mice with *Trichophyton rubrum, J. Invest. Dermatol.*, 19, 374, 1952.

56. Van Cutsem, J. and Janssen, A. J., Experimental systemic dermatophytosis, *J. Invest. Dermatol.*, 83, 26, 1984.

57. Van Cutsem, J., Fransen, J., and Janssen, P. A. J., Animal models for systemic dermatophyte and candida infection with dissemination to the skin, in *Models in Dermatology*, Maibach and Lowe, Eds., Karger, Basel, 1985, 196.

58. Van Cutsem, J., The antifungal activity of ketoconazole, *Am. J. Med.*, 74 (Suppl. B), 9, 1983.

59. Sakai, S., Saito, G., Inoue, K., nad Momoki, Y., Studies on the screening test of antifungal agent using experimental trichophytosis, *Jpn. J. Med. Mycol.*, 1, 252, 1960.

60. Tawara, K., Sunagawa, N., and Takema, M., Studies on antifungal activity of 710674-S, a new antimycotic Imidazole. II. Therapeutic effect on experimental dermatophytosis, *Jpn. J. Med. Mycol.*, 25, 351, 1984.

61. Weinstein, M. J., Oden, E. M., and Moss, E., Antifungal properties of tolnaftate *in vitro* and *in vivo, Antimicrob. Agents Chemother.*, 1964, 595, 1965.

62. Gordee, R. S. and Matthews, J. R., Evaluation of the *in vitro* and *in vivo* antifungal activity of pyrrolnitrin, *Antimicrob. Agents Chemother.*, 1967, 378, 1968.

63. Emmons, C. W., Binford, C. H., Utz, J. P., and Kwon-Chung, K. J., The mycetomas, in *Medical Mycology*, 3rd ed., Emmons, C. W., et al., Eds., Lea & Febiger, Philadelphia, 1977, 437.

64. Rippon, J. W., Mycetoma, in *Medical Mycology: The Pathogenic Fungi and the Pathogenic Actinomycetes*, 2nd ed., Rippon, J. W., Ed., W. B. Saunders, Philadelphia, 1982, 79.

65. Conant, N. F., Smith, D. T., Baker, R. D., and Callaway, J. L., Maduromycosis, in *Manual of Clinical Mycology*, 3rd ed., Conant, N. F., et al., Eds., W. B. Saunders, Philadelphia, 1971, 458.

66. Fukushiro, R., Maduromycosis, in *Handbook of Dermatology*, 7 A infections dermatosis II, Yamamura, Y., Kukita, A., Sano, S., and Seiji, M., Eds., Nakayama Shoten, Tokyo, 1982, 66.

67. Zlotnik, H. and Buckley, H. R., Experimental production of actinomycetoma in BALB/c mice, *Infect. Immun.*, 29, 1141, 1980.

68. Watanabe, S., Mycetoma in Japan, in *Proc. Primer Simp. Int. Micetomas*, Universidad Centro Occidental Microbiologia, Barquisimeto, 1978, 235.

69. Fukushiro, R., Kagawa S., Nishiyama, S., Sasagawa, S., Nishiwaki, M., Kitamura, K., Ikeda, M., Kaji, T., Yagishita, K., Hirone, T., Kumagai, T., and Matsumoto, R., *Some observations on nocardiosis, Nishinihon hifu.*, 31, 1969.

70. Destombes, P., Mariat, J., Nazimoff, O., and Satre, J., A propos des mycetomes a nocardia, *Sabouraudia*, 1, 161, 1961.

71. Conde, C., Melendro, E. I., Fresan, M., and Ortiz-Ortiz, L., *Nocardia brasiliensis* mycetoma induction and growth cycle, *Infect. Immun.*, 38, 1291, 1982.

72. Conde, C., Mancilla, R., Fresan, M., and Ortiz-Ortiz, L., Immunoglobulin and complement in tissues of mice infected with *Nocardia brasiliensis*, *Infect. Immun.*, 40, 1218, 1983.

73. Ximenez, C., Melendro, E. I., Gonzalez-Mendoza, A., Garcia, A. M., Martinez, A., and Oritiz-Oritiz, L., Resistance to *Nocardia brasiliensis* infection in mice immunized with either nocardia or BCG, *Mycopathologia*, 70, 117, 1980.

74. Beaman, B. L. and Scates, S. M., Role of L-Forms of *Nocardia caviae* in the development of chronic mycetomas in normal and immunodeficient murine models, *Infect. Immun.*, 33, 893, 1981.

75. Kurup, P. V. and Sandhu, R. S., Isolation of *Nocardia caviae* from soil and its pathogenicity for laboratory animals, *J. Bacteriol.*, 90, 822, 1965.

76. Rippon, J. W. and Peck, G. L., Experimental infection with *Streptomyces madurae* as a function of collagenase, *J. Invest. Dermatol.*, 49, 371, 1967.

77. Reiss, F., Experimental mycotic infections on laboratory animals, in *Medical Mycology*, Simons, R. D. G. P., Ed., Elsevier, London, 1954, 50.

78. Murray, I. G., Spooner, E. T. C., et al., Experimental infection of mice with *Madurella mycetomi*, *Trans. R. Soc. Trop. Med. Hyg.*, 54, 335, 1960.

79. Indira, P. U. and Sirsi, M., Studies on maduramycosis, *Indian J. Med. Res.*, 56, 1265, 1965.

80. Avram, A., Grains experimentaux maduromycosiques et actinomycosiques a *Cephalosporium falciforme*, *Monosporium apiospermum*, *Nadurella mycetomi*, et *Nocardia asteroides*, *Mycopathologia*, 32, 319, 1967.

81. Avram, A., Experimental induction of grains with *Cephalosporium falciforme*, *Sabouraudia*, 5, 89, 1945.

82. Symmers, D., Experimental reproduction of maduromycotic lesions in rabbits, *Arch. Pathol.*, 39, 358, 1945.

83. Jidoi, J. and Matsumoto, K., A case of cutaneous Aspergillosis, *Rinsho Hifu*, 20, 561, 1966.

84. Fukushiro, R., Kinbara, T., Nagai, T., Ikeda, S., and Kumagai, T., On primary pyoderma-like aspergillosis, *Jpn. J. Med. Mycol.*, 14, 127, 1973.

85. Burgoon, C. F., Blank, F., Johnson, W. C., and Grappel, S. F., Mycetoma formation in Trichophyton rubrum infection, *Brit. J. Dermatol.*, 90, 155, 1974.

86. Baylet, R., Camain, R., Juminer, B., and Faye, I., *Microsporum ferrugineum* ota, 1921, agent de mycetomes du cuir chevelu en Afrique noire, *Pathol. Biol.* (Paris), 21, 5, 1973.

87. Frey, D. and Lewis, M. B., Mycetoma of the scalp in an aboriginal child, *Aust. J. Dermatol.*, 17, 7, 1976.

88. Howard, D. H. and Orr, G. F., Comparison of strains of *Sporotrichum schenckii* isolated from nature, *J. Bacteriol.*, 85, 816, 1963.

89. Nakahara, T., Studies on the strains of *Sporotrichum schenckii* isolated from soils, *Jpn. J. Med. Mycol.*, 12, 30, 1971.

90. Mariat, F., Adaptation de *Ceratocystis* à la vie parasitaire chez l'animal — Etude de l'aquisition d'un pouvoir pathogene comparable à celui de *Sporothrix schenckii*, *Sabouraudia*, 9, 191, 1971.

91. Watanabe, S., Tropical heat therapy for sporotrichosis, in *Proc. XVIth Int. Congr. Dermatol.*, Kukita, A. and Seiji, M., Eds., University of Tokyo Press, Tokyo, 1983, 341.

92. Mackinnon, J. E., The dependence on the weather of the incidence of sporotrichosis, *Mycopathologia*, 4, 367, 1949.

93. Mackinnon, J. E. and Conti-Diaz, The effect of temperature on sporotrichosis, *Sabouraudia*, 2, 56, 1962.

94. Rippon, J. W., Ed., *Medical Mycology: The Pathogenic Fungi and the Pathogenic Actinomycetes*, 2nd ed., W. B. Saunders, Philadelphia, 1982, 291.

95. Satterwhite, T. K., Kageler, W. V., Conkoin, R. H., Portnoy, B. L., and DuPont, H. L., Disseminated sporotrichosis, *J. Am. Med. Assoc.*, 240, 771, 1978.

96. Kwon-Chung, K. J., Comparison of isolates of *Sporothrix schenckii* obtained from fixed cutaneous lesions with isolates from other types of lesions, *J. Infect. Dis.*, 139, 424, 1979.

97. Charoenvit, Y. and Taylor, L. R., Experimental sporotrichosis in Syrian hamsters, *Infect. Immun.*, 23, 366, 1979.

98. Barbee, W. C., Ewert, A., and Davidson, M., Animal model: sporotrichosis in the domestic cat, *Am. J. Pathol.*, 86, 281, 1977.

99. Okudaira, M., Tubura, E., and Schwarz, J., A histopathological study of experimental murine sporotrichosis, *Mycopathologia*, 14, 284, 1961.

100. Taylor, J. J., A comparison of some ceratocystis species with *Sporothrix schenckii*, *Mycopath. Mycol. Appl.*, 42, 233, 1970.

101. Baker, R. D., Experimental sporotrichosis in mice, *Am. J. Trop. Med.*, 27, 749, 1947.

102. Benham, R. W. and Kesten, B., Sporotrichosis: its transmission to plants and animals, *J. Infect. Dis.*, 50, 437, 1932.

103. Kobayashi, T., Beitrage zur experimentellen Sporotrichose. I. Mitteilung: Impfversuche mit *Sp. Beurmanni* (Stamm von Kobayashi) an weissen Ratten und Kaninchen, *Jpn. J. Dermat.*, (in Japanese), 38, 747, 1935.
104. Kinbara, T. and Fukushiro, R., Fungal elements in tissue, in sporotrichosis, in Proc. XVIth Int. Congr. Dermatol., Kukita, A. and Seiji, M., Eds., University of Tokyo Press, Tokyo, 1983, 348.
105. Staib, F., Randhawa, H. S., and Blisse, A., Observations on experimental sporotrichosis of the muscle, *Bakt. Hyg., I. Abt. Orig.*, A221, 250, 1972.
106. Hopkins, J. G. and Benham, R. W., Sporotrichosis in New York State, *N.Y. State J. Med.*, 32, 595, 1932.
107. Miyaji, M. and Nishimura, K., Defensive role of granuloma against *Sporothrix schenckii* infection, *Mycopathologia*, 80, 117, 1982.
108. Shiraishi, A., Nakagaki, K., and Arai, T., Experimental Sporotrichosis in congenitally athymic (nude) mice, *J. Reticuloendothel. Soc.*, 26, 333, 1979.
109. Reiss, F., Experimental mycotic infections on laboratory animals, in *Medical Mycology*, Simons, R. D. G. P., Elsevier, London, 1954, 50.
110. Hektoen, L. and Perkins, C. F., Refractory subcutaneous abscesses caused by *Sporothrix schenckii*, a new pathogenic fungus, *J. Exp. Med.*, 5, 77, 1900.
111. Kren, O. and Schramek, M., Ueber Sporotrichose, *Wien. Klin. Wochenschr.*, 22, 1519, 1909.
112. Warfield, L. M., Report of a case of disseminated gummatous sporotrichosis, with lung metastasis, *Am. J. Med. Sci.*, 164, 72, 1922.
113. Büchel, K. H., Plempel, M., and Bartmann, K., Experimental study on the chemistry and antifungal activity of Clotrimazole, *Medizin von Heute (Bayer)*, (in Japanese), 60, 76, 1976.

Chapter 3

ANIMAL MODELS OF SYSTEMIC MYCOSES

E. Brummer and K. V. Clemons

TABLE OF CONTENTS

I. BLASTOMYCOSIS

A. Introduction

Blastomyces dermatitidis, a thermally dimorphic fungus, is a pulmonary pathogen in humans,[1-4] dogs,[5,6] mice,[7,8] and other mammals.[7] Untreated disseminated-human blastomycosis has been reported to have a mortality rate greater than 80%.[9] However, virulent and avirulent isolates of *B. dermatitidis* have been described[8,10,11] and in five outbreaks, blastomycosis in some patients (28 of 46) resolved without therapy.[4] Blastomycosis is naturally acquired by way of the pulmonary route[12] by inhalation of the infectious agent in its saprophytic or mycelial-conidia phase. Five documented point outbreaks, summarized by Sarosi et al.[3] and Tenenbaum et al.,[4] have been associated with construction activity and postulated disturbance of *B. dermatitidis* in its ephemeral ecological niche.[13-16] Although the dog was the first animal used in the study of blastomycosis[17] and in later epidemiological studies,[5,6] the laboratory mouse has been the animal of choice for isolation of *B. dermatitidis* from nature[13,14] and development of an animal model of pulmonary blastomycosis.[8] The murine model of pulmonary blastomycosis has been used extensively over the past several years for immunological[18-21] and drug treatment studies,[22,23] consequently it will be described first and in detail.

B. Murine Model of Pulmonary Blastomycosis

1. Introduction

A murine model of pulmonary blastomycosis was developed by Harvey et al.[8] in 1979. In this model, pulmonary infection was established by intranasal instillation of the yeast-form, resulting in its aspiration into the lungs by the lightly diethyl ether-anesthetized mouse. With this procedure, 33.5% of the inoculum colony-forming units (CFU) was recoverable from the lungs less than 1 hr after infection.[18] The severity of infection and extent of mortality were dependent on the isolate and dose of *B. dermatitidis*,[8,18,19] as well as maturity[18,19] and strain of mouse.[18]

2. Blastomyces dermatitidis (Yeast Form)

Isolates of *B. dermatitidis* used in this model can be obtained from American Type Culture Collection (ATCC), Rockville, Maryland. ATCC 26199 (SCB-2, a human isolate) and ATCC 10285 (unspecified origin) are the most virulent in mice[3] compared to other isolates, e.g., ATCC 26198 (KL-1, a soil isolate) and ATCC 26197 (GA-1, human cutaneous isolate). Cultures grown on brain-heart infusion (BHI) agar slants at 37°C, 4 to 5 days, can be stored under water at 4°C for 3 to 6 months. Alternately, cultures grown in broth (synthetic amino acid medium for fungi, SAAMF)[24] at 37°C can be stored in small aliquots at −70°C. Repeated transfer of ATCC 26199 on BHI weekly for 1 year,[18] with less frequent transfers (every 3 to 6 months), followed by storage at 4°C under water over 4 to 5 years resulted in attenuation of virulence.[146] Virulence of ATCC 26199 was not restored by up to five animal passages in mice.[146] ATCC 26199 is stored by ATCC in the lyophilized state and has retained virulence since its deposit. Inocula of *B. dermatitidis* for pulmonary infection are grown by plating 0.5 mℓ of growth in SAAMF (37°C, gyratory shaker, 72 to 96 hr) on a blood-agar plate (37°C, 72 hr). Pasty growth is harvested with a wire loop, washed twice in saline, counted in a hemacytometer and 1 mℓ of appropriate dilutions plated on blood-agar plates in triplicate to determine CFU per mℓ. The ratio of CFU to hemacytometer counts of units (single and multicellular) is usually 0.8. For intranasal inoculation, mice are lightly anesthetized with diethyl ether until the eye-blink reflex (also sneeze and swallow reflexes) is lost. The mouse is held in the palm of the hand with the index finger behind the head, the thumb closing the mouth, then 0.03 mℓ of inoculum is slowly applied to the tip of the nose using an adjustable 0.05 mℓ microdiluter. This allows the inoculum to be aspirated into the respiratory tract.

3. Mouse Strains and Age

An advantage of the murine model of pulmonary blastomycosis is the availability of inbred strains. Although the majority of published works[8,19,20,22,23] have utilized BALB/c male mice, other strains have been used to address specific questions, e.g., C3H/HeJ mice vs. C3H/HeN as well as A/HeJ, C3HSW/sn, SLJ/J, C57Bl/10J, DBA/1J, and DBA/2J.[18] The maturity of mice has been shown to be a critical factor in resistance to pulmonary infection[19] regardless of the strain of mouse.[18] For example, the LD$_{50}$ for mice less than 15 g (approximately 4 to 5 weeks of age) was 1000 times less than for 25 g (8- to 10-weeks-old) BALB/c male mice and ATCC 26199 attenuated strain.[19]

4. Histopathology

The histopathology of murine pulmonary blastomycosis has been described by Harvey et al.[8] and confirmed by Brass and Stevens.[19] Briefly, little host-parasite interaction was evident at 2 days post-infection, however at 4 to 6 days, 10 to 20% of the lung parenchyma had focal pneumonic processes consisting of *B. dermatitidis*, polymorphonuclear leukocytes, and mononuclear cells. This condition progressed until 70 to 80% of the lung parenchyma was involved at 10 to 12 days of infection. After 15 days of infection, lungs were covered with necrotic white nodules and 95% of the tissue was consolidated in a confluent, but clearly distinguishable, pneumonic infiltrate. In this rather acute type of pulmonary infection, granuloma formation was not observed. Dissemination of *B. dermatitidis* from the lungs to other organs, e.g., liver and spleen, was documented by Harvey et al.[8] in 64% of the mice infected with less than 100 CFU, rather than higher doses, as was associated with protracted disease.

5. Applications

Development of the murine model of pulmonary blastomycosis has made several different types of studies possible. Valuable and pragmatic use has been made of this model in testing the efficacy of new drugs, e.g., imidazoles such as ketoconazole, compared to amphotericin B,[22] for treatment of this disease. This system has also been utilized to evaluate the potency of immunomodulators, e.g., muramyl dipeptides (MDP), in enhancing nonspecific resistance to a pulmonary fungal infection.[23] In immunological studies, mice immunized by resolution of a subcutaneous infection were challenged as in the murine model of pulmonary blastomycosis to demonstrate resistance.[20] Similarly, the model was used to demonstrate the adoptive transfer of resistance with splenic T-lymphocytes from immunized mice.[21] This model continues to be used as a probe for virulence factors in *B. dermatitidis*, for example virulent (ATCC 26199) vs. avirulent (ATCC 26198) isolates[8,19,25] and host defenses in young vs. adult mice.[19]

C. Intraperitoneal, Subcutaneous, and Intravenous Murine Models of Infection

1. Introduction

Simplicity and convenience of the peritoneal route of experimental infection with *B. dermatitidis* has favored its use from as early as 1906[26] until the present.[28] For these same reasons, and the advantages of quantitative recovery, the subcutaneous abscess model of *B. dermatitidis* infection has also been used extensively.[11,18,20,21,29-31] The intravenous route of infection has been used primarily in the isolation of *B. dermatitidis* from nature.[13,14]

2. Intraperitoneal Route

The earliest experiments with *B. dermatitidis* in mice were done by Bowen and Wolbach[26] using the peritoneal route in 1906 some 80 years ago, In 1942, Baker[27] also used this model to make detailed histopathological observations of host-parasite interactions. More recently, this route of infection has been employed to assess protection afforded by vaccination,[32] by resolution of subcutaneous infection,[20,28] by maturity,[19,25] or by inbred strain of mice.[33] In general, the yeast-form of *B. dermatitidis*, grown at 30°C on BHI agar[28,32] or blood agar,[19,25,33] was injected intraperitoneally, then mortality was recorded,[19,20,25,28,33] or dissemination to lung, liver, and spleen was measured as recoverable CFU.[32]

3. Subcutaneous Route

When mice were infected subcutaneously with doses of *B. dermatitidis* (ATCC 26199), up to 40,000 CFU, the infection was nonlethal[20] the infection was walled off in an abscess and sterilized in 4 weeks.[11,20] Resolution of such infections resulted in a profound resistance to lethal pulmonary or intraperitoneal challenges with *B. dermatitidis*[20,21] and accelerated clearance of *B. dermatitidis* from sites on reinfection.[20] A technical advantage of this system is the ability to accurately deliver a fixed number of CFU to a site where they remain and can be retrieved at various times later. This feature has been exploited to show that virulence of *B. dermatitidis* correlates with ability to replicate in vivo.[11] It also has been used to demonstrate the effect of polymorphonuclear neutrophils (PMN) on *B. dermatitidis* replication in vivo by coinjection of PMN and *B. dermatitidis* subcutaneously in a Winn-type assay.[30] Subcutaneous abscesses are visible by Day 4, well-defined by Day 7, can be easily dissected *in toto*, processed with a tissue grinder, and CFU per site quantitated by plating on blood-agar plates.[11,30]

4. Intravenous Route

Denton and DiSalvo used the intravenous route in their method for isolation of *B. dermatitidis* from soil[13,14] and also employed this system in subsequent studies of virulence of *B. dermatitidis* isolates for mice.[10] This may be a more sensitive test of virulence of *B. dermatitidis* isolates because of efficient delivery of the inoculum to susceptible organs. For example, isolates KL-1 (ATCC 26198) and SL-1 were 80 to 100% lethal when given intravenously,[10] whereas KL-1 had moderate virulence in pulmonary challenge[8] and in its ability to replicate in subcutaneous sites[146] compared to ATCC 26199. The intravenous route has also been used to test the resistance of immunized mice as measured by their ability to restrict replication of *B. dermatitidis* in spleen, liver, and lungs.[32]

D. Mycelial-Conidia Infection Models of Blastomycosis

1. Introduction

As described above, most models of blastomycosis used the parasite yeast-form for infection. This by-passes the first events that take place in natural infections, namely, conversion from the saprophytic to the parasitic form in the host. This short cut to infection served certain purposes well but has left many questions unanswered relative to host-parasite interaction in the more natural type of infection with *B. dermatitidis*. Infectivity of mycelial vs. yeast forms,[34] effect of sex hormones on resistance to infection with the mycelial form,[35] and relative susceptibility of five kinds of laboratory animals to infection[36] represent most of the published work in this interesting area.

2. Blastomyces dermatitidis (Mycelial-Conidia)

Human and animal yeast-form isolates were shown to grow in the mycelial form on standard mycological media when incubated at room temperatures; however, the mycelial growth consisted mainly of mycelial elements and some microconidia which were difficult to quantitate for experimental purposes.[37] Smith and Furcolow[37] and Smith[38] showed that microconidia formation was greatly enhanced when *B. dermatitidis* was grown and passaged five times on soil-infusion or starling (*Sturnis vulgaris*) manure-infusion agar medium,[37] or more conveniently, on yeast extract (Difco) agar.[38] The latter consisted of 8 mℓ of filtered (0.45 μm) 15% yeast extract added to 1 ℓ of auto-claved agar (2%) cooled to 45°C. Forty mℓ of medium in 250 flasks were inoculated and incubated at 25°C for 28 days, then the growth was harvested by gently shaking after adding 10 ml of distilled water and glass beads. A canine isolate grown in this fashion yielded 92% microconidia (3 μm) and 8% hyphal elements with 85% viability, whereas a human isolate produced 87% microconidia and 13% hyphal elements with 55% viability.[38]

3. Mice

Murine models of blastomycosis established by infection with microconidia are few in number and elementary; however, valuable information has been gained from them. Baker, in 1938, showed that the mycelial form was just as infective as the yeast form, on a weight basis, when given intraperitoneally to young white mice (15 to 21 g) of unspecified sex.[34] More quantitative studies were done by C. D. Smith et al.[36] in which microconidia (144,000 viable units) were given intraperitoneally to young (weanling) male Swiss mice. *B. dermatitidis* was recovered in 4 of 10 mice when the lower right lobe of the lung, 1 g of liver, and the spleen were macerated and cultured on Sabouraud's dextrose agar. The only other published study using microconidia to infect mice was the important experiments of Denton and DiSalvo[7] in which they demonstrated the respiratory infection of mice with *B. dermatitidis* subsequent to inhabitation of cages containing soil inoculated with microconidia.

4. Hamsters

The hamster can be infected with *B. dermatitidis*[39] and appears to be much more susceptible to infection with microconidia of *B. dermatitidis* than mice.[35] When 144,000 viable microconidia were given intraperitoneally to weanling male hamsters, which are several fold larger than mice, 10 of 10 had recoverable *B. dermatitidis* after 1 month compared to 4 of 10 weanling male Swiss mice. The greater susceptibility of hamsters was not isolate-dependent because 4 isolates, two human and two canine, demonstrated this phenomenon. The rat (17%) and guinea pig (20%) were susceptible to this inoculum; however, their susceptibility could not be compared to that of the hamster because the inoculum size was not adjusted to the weight of the animals.[35] These findings suggest that hamsters may be more efficient animals than mice in isolation of *B. dermatitidis* from nature using the method of Denton et al.[13] One of the most provocative and elegant studies with hamsters were those of Landay et al.[35] on the effect of sex on susceptibility. Females were shown to be significantly more resistant (52% survival) than males (7% survival). Castrated males were more resistant than normal males (40% vs. 7% survival) and treatment of ovariectomized females with testosterone rendered them as susceptible as males (47% vs. 52% survival). These findings provide the basis for further investigations to explain the greater ratio (9:1) of male to female cases of human blastomycosis reported by Conant et al.[40]

II. COCCIDIOIDOMYCOSIS

A. Introduction

Coccidioidomycosis is a mycotic disease caused by the dimorphic fungus *Coccidioides immitis*. Historically, the disease has been recognized for less than a century. The first documented cases were reported in the late 1800s by Posadas,[41] Wernicke,[42] Rixford,[43] and Rixford and Gilchrist.[44] These authors accurately described the disease; however, they considered the mold growing from bacteriologic specimens to be a contaminant and the organism observed in tissue specimens to be a protozoan. Only a few years later, Ophüls and Moffitt[45] discovered that the mold, previously discarded as a contaminant, was the etiologic agent of the disease. In the years that have followed, a great deal of effort has been made by numerous investigators to document and to understand the fungus and its ecology, the epidemiology of the disease, and the various aspects of the host-parasite interactions including host response, treatment, and prevention. Several excellent sources are available which review and discuss the many facets of coccidioidomycosis.[46-49]

C. immitis has been found only in the arid and semi-arid soils of the southwestern

United States and northern Mexico, as well as parts of Central and South America.[49] The disease arises after inhalation of the saprobic form arthroconidia from the soil. In host tissues, these arthroconidia convert into the parasitic form of the organism, endosporulating spherules.[49] The disease presents a variety of clinical manifestations which range from asymptomatic primary pulmonary infection in approximately 60% of the cases, to more severe, acute, or chronic disseminated mycosis.[46,47,49] Natural infections of man are common in the endemic areas of the United States and become a problem because of the associated morbidity.[49] Recovery, with concomitant development of a strong cell-mediated immune response, usually confers a high degree of resistance;[47,49] however, predisposition to the more serious and life-threatening forms of coccidioidomycosis have been associated with various factors such as race, age, sex (hormonal), pregnancy, and overall immune status.[49,50] In addition to man, many species of mammals also are susceptible to infection by *C. immitis*. Some of the species include dogs, cats, primates, horses, cattle, swine, sheep, rodents, and even a sea otter.[47,49] Of these, primates, certain breeds of dogs, and rodents are probably the most susceptible with swine, cattle, and cats being the least susceptible.[47,49] Much of the present day understanding about *C. immitis*, and the various disease states it causes, has been accumulated by the development of animal models of experimental infection used to study the course of disease, host response, treatment, and prevention. While a number of different animal models have been described (i.e., rat, guinea pig, rabbit, dog, primate), the laboratory mouse is the most often used and will be described in the most detail.

B. Murine Models of Coccidioidomycosis
1. Introduction

The laboratory mouse has been used in numerous studies on *C. immitis*. These include studies on the immunology, pathology, antifungal therapy, and immunoprophylaxis of coccidioidomycosis. Most commonly, the arthroconidia are instilled or inoculated by the desired route of infection. Low doses of arthroconidia (i.e., <50 to 100) often result in chronic infections whereas increased doses (i.e., >100 to 500) result in rapidly fatal disease. While strains of *C. immitis* vary in virulence, selection of a suitable strain and careful maintenance of cultural conditions allow for a predictable experimental disease. *C. immitis* grows readily on common mycological media at 25°C to 35°C. Stock cultures of *C. immitis* cultured at 25°C on slants of glucose-yeast extract agar or Sabouraud's dextrose agar grow rapidly with the initiation of arthroconidia formation after 7 to 10 days. Abundant arthroconidia are present within 2 to 3 weeks of culture. A note of caution in that all *C. immitis* cultures should be handled in only the most rigorous of containment facilities by experienced personnel because of the extreme danger of arthroconidial dispersion by air currents. Single-cell conidia suspensions can be prepared by several methods and enumerated by hemacytometer and viable count.[51,52] Arthroconidia can be stored in distilled water at 4°C until needed.[52] Spherules can be isolated from heavily infected animal tissues or grown in vitro in modified Converse medium.[53,54]

2. Pulmonary Infection

Intranasal (i.n.) instillation of arthroconidia is used most often to produce pulmonary disease. Animals are first anesthetized with ether or an ether, chloroform, ethanol solution (3:2:1)[55] or by intraperitoneal injection of pentabarbital sodium (0.06 mg/g body weight)[56] and the desired number of arthroconidia instilled by placing 0.05 ml on the nostrils to be inhaled by the animal. Given by the i.n. route, approximately 50 arthroconidia represent one LD_{50}.[57] The cellular responses after i.n. instillation of arthroconidia have been detailed thoroughly by several authors and are consistent with a progressive mixed granuloma composed of mononuclear and granulocytic cell types.[58-61] In nonimmune mice, *C. immitis* converts to the spherule phase within 48 to 96 hours and multiplies rapidly, causing a more purulent exudate with abundant spherules;[60,61] extrapulmonary dissemination has been reported with doses of 200 arthroconidia by the 14th day of infection.[56] Immunized mice are better able to restrict extrapulmonary dissemination,[58,59] exhibiting more compact granulomata by 2 to 4 days postinfection.[59,60] The severity of pulmonary disease is also reflected by an increase in lung weight[58,60] during progressive disease as well as an increase in recoverable *C. immitis* as determined by plate counts.[56,58,59]

The pulmonary model of coccidioidomycosis provides a rigorous challenge route simulating natural infection and has been used to evaluate prospective antifungal therapies,[51,55,62] vaccine preparations,[57-60,63,64] and host response to *C. immitis*.[56,58-60]

3. Systemic Coccidioidomycosis

a. Introduction

Systemic coccidioidomycosis can be produced in mice by injection of *C. immitis* either intraperitoneally (i.p.) or intravenously (i.v.) The i.p. route provides a rapid method of inoculation with the organism recoverable from the visceral organs (i.e., spleen, liver omentum) and eventual dissemination to the lungs.[62,65] However, i.p. inoculation of *C. immitis* is less rigorous than is either i.n. or i.v.[57,66]

b. Intraperitoneal Route

Several studies utilized the i.p. route to study resistance mechanisms and potential vaccines against *C. immitis*.[57,63-65,67-69] These studies have used various preparations of *C. immitis* as potential immunogens, demonstrating enhanced resistance to challenges containing up to several thousand arthroconidia; however, similarly treated animals challenged i.n. were observed to be significantly less resistant.[57,67] Beaman et al.[68,69] used i.p. challenged animals to examine the significance of T-cells during coccidioidomycosis. Their findings include increased susceptibility to *C. immitis* after adult thymectomy or by congenitally athymic animals as well as the capacity of T-cells to adoptively transfer resistance.[68,69] More recently the i.p. model of coccidioidomycosis has been used to examine the genetic basis of resistance to this disease.[65] Using genetically defined mice, Kirkland and Fierer demonstrated that DBA/2NX1 ($\log_{10}$ LD$_{50}$ = 5.25) mice were approximately 1000-fold more resistant to *C. immitis* than were BALB/cAnN mice ($\log_{10}$ LD$_{50}$ = 1.67).[65] In addition, C57BL/10N (LD$_{50}$ = 2.77) and C57BL/6N (LD$_{50}$ = 2.83) were more resistant than were BALB/c, but less resistant than were DBA/2NX1 mice.[65] Challenge of F1 hybrids demonstrated resistance to be the dominant phenotype.[65] It is important to note that these results may be valid only for i.p. challenged animals, since Beaman et al.[68] have reported that DBA/2 mice can be infected i.n. with relatively few arthroconidia. In addition, Clemons et al.[56] have reported that C57BL/6 Bg/+ and Bg/Bg mice were as susceptible to i.v. challenge as were BALB/c nu/nu mice (LD$_{50}$ < 10 arthroconidia) and at least 6-fold more susceptible than BALB/c nu/+ mice (LD$_{50}$ = 75 arthroconidia). These somewhat conflicting results indicate the possibility that resistance may depend upon the route of challenge as well as genetic background. Similar results have been reported for *Blastomyces dermatitidus*.[70] Clearly, comparative studies are needed to resolve this conflict.

c. Intravenous Route

Intravenous inoculation of *C. immitis* is less frequently used to establish systemic disease. Sinski and Soto[61] injected massive doses of *C. immitis* (>10^7 arthroconidia) i.v. in order to follow the early stages of cellular response in the lungs. Hyphal fragments and cellular changes were observed as early as 6 hr after challenge and spherules were apparent within 48 hr.[61] Scalarone and Huntington[66] have reported the relative avirulence of spherule-endospore *C. immitis* given i.v. as compared to mycelial phase arthroconidia. In addition, they were able to produce a high incidence of circling syndrome, attributable to inner ear disease, using i.p. inoculated endospores but not i.v. inoculated spores.[66] Clemons et al.[56] used i.v. inoculation of *C. immitis* arthroconidia to study resistance mechanisms in immunodeficient murine models. In those studies age-related, but not sex-related, as well as genetic background related differences in susceptibility were observed.[56] These differences were apparent in mortality and as progressive increases in tissue burdens of *C. immitis*.[56] Histologically, i.v. inoculation of *C. immitis* resulted in a normal mixed-cellular response, except in BALB/c nude mice which exhibited only a suppurative response and C57BL/6 beige mice which exhibited primarily a mononuclear response.[56,61]

C. Other Animal Models of Coccidioidomycosis
1. Introduction
Several other animals (i.e., dogs, guinea pigs, and monkeys) have been used to study possible vaccines against *C. immitis* as well as immunologic responses.[52,71-73] Because a humoral response to *C. immitis* is usually not detectable in mice, the use of other animal models can provide experimental results that more closely parallel the course of disease in humans.

2. Monkeys
Levine et al.[73] used cynomologous monkeys to study the effectiveness of a killed spherule vaccine. Their results indicated that approximately 5 mg/kg provided protection against an otherwise lethal aerosol of 200 arthroconidia.

Lowe et al.[74] examined the serologic and skin test responses of *Macaca mulatta* monkeys after vaccination and aerosol challenge with *C. immitis*. Their results indicated similarities to human responses with some discrepancies in the tube-precipitin reactions from the monkeys sera.[74] Other investigators also have described experimental models of coccidioidomycosis using monkeys.[71,75,76]

3. Guinea Pigs
Cox et al.[52,72] infected guinea pigs intratracheally with 50 arthroconidia of *C. immitis* and followed the temporal course of disease culturally and immunologically. They found the overall immunologic responses of guinea pigs to infection were consistent with the pattern in humans but with no obvious correlation between dissemination and complement-fixing antibody titers.[52] In addition, extrapulmonary dissemination of *C. immitis* was found to be an early event in guinea pigs.[52] Other investigators have demonstrated guinea pigs were susceptible to aerosol infection.[77] Guinea pigs also have been used to study responsiveness to *C. immitis* antigens with the induction of anergy,[78] skin-test responsiveness,[52,72,78,79] and adoptive transfer of skin-test sensitivity utilizing transfer factor.[80]

III. HISTOPLASMOSIS

A. Introduction
Histoplasma capsulatum, a thermally dimorphic fungus, is a pulmonary pathogen in humans and the causative agent of Darling's disease[81] or histoplasmosis.[82] Unlike blastomycosis and paracoccidioidomycosis, where the true nature of the causative

agents were identified in conjunction with description of the disease, some 28 years elapsed (1906 to 1934) before deMonbreun cultured, isolated, and established the dimorphic nature of *H. capsulatum.*[82] *H. capsulatum* was isolated from soil in 1949,[83] as well as soil contaminated with bird dung[84] or bat guano.[85] The spread of endemic areas in the United States paralleled the importation of the European starling (*Sturnus vulgaris*) in 1896 and its spread throughout the Ohio-Mississippi River Valley by 1936.[86] Skin testing with the histoplasmin of Van Pernis by Furcolow[87] showed that the largest area of high prevalence, an 80 to 90% positive skin-test population, was the Ohio-Missouri-Mississippi River Valley of the U.S.; however, scattered areas with high prevalence exist in all parts of the world. Although a high rate of infection occurs in endemic areas (80 to 90% positive skin test), only a few individuals develop chronic progressive disease. Epidemics of acute disease, usually associated with human activity involving disturbance of heavily contaminated bird or chicken roosts, have been recorded.[88] Natural infections in dogs[89] and other domestic animals, e.g., cats, horses, etc.[90] have been reported. However, only bats harbor the disease and are an important vector for dissemination,[91] whereas, chickens and birds are relatively resistant to infection.[92] Of laboratory animals, the mouse is the most susceptible,[93] and murine models of experimental histoplasmosis have been developed and are described in detail below.

B. Murine Models of Histoplasmosis
1. Pulmonary Histoplasmosis
a. Introduction

Although natural infection with *H. capsulatum* occurs by inhalation of microconidia (microaleuriospores) or macroconidia, we are aware of only one published study of pulmonary infection in mice using this form of *H. capsulatum.*[94] However, two reports have appeared in which mice were challenged intratracheally with the yeast-form of *H. capsulatum.*[95,96]

b. Histoplasma capsulatum (Isolates)

The yeast form of human isolates of *H. capsulatum* are grown as stock cultures on BHI agar supplemented with 1% glucose and 0.1% L-cysteine hydrochloride at 37°C then stored at 4°C for 4 to 6 weeks.[96]

For animal inocula preparations, BHI broth supplemented with 1% glucose (250 m*l* in a 1 *l* flask) was seeded with a 15 m*l* BHI broth-yeast cell suspension and incubated at 37°C on a gyratory shaker (150 rpm) for 36 hr.[97] Washed yeast cells are counted with a hemacytometer, and viability maybe measured by the conversion of yeast to mycelial-form by the method of Sun and Huppert[98] or more easily by the Janus green B vital-staining method of Berlinger and Reca.[99] Plating efficiency of *H. capsulatum* is very poor (<20%) on standard mycological media but can be greatly improved by the addition of spent media or a siderophore, hydroxamic acid, present in spent media.[100]

Soil isolates of *H. capsulatum* can be grown on simple cornmeal agar at room temperature.[94] Six-week-old cultures yielded numerous tuberculate macroconidia (5 μ when harvested with a wire loop under sterile saline and were 60 to 80% viable when grown on cornmeal agar.[94] Others have obtained mycelial-forms of *H. capsulatum* for inocula preparation by conversion of yeast to mycelial-forms on mycosel agar (BBL, Cockeysville, Md.) for 3 to 4 weeks at 28°C. This growth was used to inoculate flasks containing Smith sporulation agar,[101] and after 5 weeks at 28°C, growth was harvested.[101] Mycelial growth was processed in a Waring blender (6, 10-sec intervals), filtered through 16-ply gauze, and centrifuged (10 × g, 10 min), yielding mostly microconidia (2 to 3 μ), mycelial elements, and very few tuberculate macroconidia.[101]

c. Pulmonary Infection

Intratracheal (i.t.) injection of 10^5 yeast cells of *H. capsulatum* established a self-resolving infection 28 days postinoculation.[95] Dissemination to the spleen was not observed in this model.[95] Bronchoalveolar lavage (BAL) of infected mice revealed that PMN increased dramatically 7 days postinfection, making up to 20 to 30% of the cells obtained by BAL. The number of lymphocytes recovered by BAL was also significantly greater in infected mice than in controls (15 vs. 2.6%). This self-resolving pulmonary infection is interesting because it probably represents the most frequent type of subclinical *H. capsulatum* infection in humans. Perhaps a higher i.t. dose, e.g. 10^6 yeast cells used for lethal challenge i.v., would produce a disseminated systemic type of infection.

2. Systemic Histoplasmosis
a. Introduction

Although systemic histoplasmosis has been produced in mice by pulmonary infection,[94] the most frequently used murine model of histoplasmosis has been the systemic type where infection was established by the intravenous route. The rationale is that this represents the most serious and clinical form of human histoplasmosis, consequently vaccination[97,102,103] and immunological[104-106] studies in such a murine model would be relevant to disseminated disease in humans.

b. Pulmonary Route

Systemic histoplasmosis resulting from pulmonary infection of mice with mycelial-forms (mainly tuberculate macroconidia) of a soil isolate has been described by Procknow et al.[94] Sequential histopathology in the lungs illustrated implantation, macroconidia rupture, and conversion to the parasitic yeast-form. Dissemination to the spleen and liver occurred as early as within 4 to 6 days, and after 10 days, these organs were heavily infected. Five to 20,000 macroconidia, 60 to 80% viable, given by intranasal instillation to ether-anesthetized mice, produced a disseminated fatal histoplasmosis. Some factors which may be important in this model are (1) small macroconidia (5 μm) produced by this isolate, (2) the growth on cornmeal agar, and (3) the age of the mice (15 g, 3 to 4 weeks of age).

c. Intravenous Route

Infection with 10^6 yeast-form cells of *H. capsulatum* by the i.v. route has proven to be a reproducible method for producing lethal systemic histoplasmosis.[97,103,106] On the other hand, Howard et al.[102] have used sublethal (2.15×10^5) doses of the yeast-form (strain 505) i.v. for establishment of profound resistance to rechallenge with lethal doses i.v. Splenic lymphocytes and peritoneal macrophages from such immunized mice have been employed extensively for immunological studies, e.g., macrophage activation.[102,105]

d. Intraperitoneal Route

Infection of mice i.p. with yeast-form *H. capsulatum* was effective in establishing systemic histoplasmosis in athymic (nu/nu) mice.[107,108] To produce similar disseminated disease, a tenfold higher dose was required i.p., compared to i.v.[108] In athymic (nu/nu) nude mice, doses (10^5, i.p.) that were sublethal in thymus-intact mice (nu/+), were 100% lethal in 50 days.[107] Even higher doses, 3×10^7 and 10^8 of some *H. capsulatum* isolates i.p., produced little (1/20) mortality in (nu/+) mice. This route of infection appears to be the one of choice for studies of systemic histoplasmosis in athymic nude mice.

e. Subcutaneous Route

Systemic histoplasmosis apparently can occur when large doses of yeast-form *H. capsulatum* are given subcutaneously (s.c.).[97,102] Tewari et al.[97] have used sublethal doses (2×10^5) s.c. for immunization of mice against rechallenge with lethal doses (10^6) i.v. and for immunological studies.[97,104] Therefore, the main use of infection by the s.c. route has been for immunization purposes.

C. Other Animal Models of Histoplasmosis
1. Introduction

Dogs,[94,109] guinea pigs,[101,110] hamsters,[110] and rabbits[111] have been used as models of systemic histoplasmosis. The mycelial-form of *H. capsulatum* was used in some of these studies and infection was established by inhalation.[94,101,109]

2. Pulmonary Route (Dogs, Guinea Pigs)

Turner et al.[109] inoculated sterilized soil with yeast-form *H. capsulatum* and incubated it at 28°C for 6 months. This was used to expose litters of suckling pups along with dams.[109] Fifty percent of the pups (18/36) died in 6 weeks and 95% in 17 weeks; whereas, none of the dams died and only 9 of 24 became *H. capsulatum* positive. In more quantitated experiments, Schlitzer et al.[101] infected guinea pigs with microconidia using a Collison nebulizer connected to a Henderson apparatus. With 1,748 viable units, an acute nonlethal infection with dissemination only to bronchiotracheal and cervical lymph nodes was established in 100% of the guinea pigs 3 weeks post infection. Resolution of such infections by 8 weeks was accompanied by cutaneous-delayed hypersensitivity reaction (9 of 11) and complement-fixing antibody titer (1:128) to *H. capsulatum* antigens.[101]

3. Intraperitoneal Route (Hamsters, Guinea Pigs)

Salvin injected mice, hamsters, and guinea pigs i.p. with yeast-form *H. capsulatum* to compare their relative susceptibility.[110] Guinea pigs cleared high doses (10^4 to 10^6) of *H. capsulatum* by 12 weeks. Hamsters were more susceptible than mice to 10^3 yeasts, e.g. 93% (11/12) were culture positive at 11 weeks vs. 13% of mice.

4. Intravenous Route (Rabbits)

Rabbits were susceptible to yeast-form *H. capsulatum* given i.v.[111] Yeasts produced from different filamentous types of *H. capsulatum* differed in virulence, as assessed by organ recovery, when injected i.v. into rabbits.

IV. PARACOCCIDIOIDOMYCOSIS

A. Introduction

Paracoccidioides brasiliensis, a thermally dimorphic fungus and the causative agent of paracoccidioidomycosis (South American blastomycosis) was isolated from patients in Sao Paulo, Brazil by A. Lutz in 1908.[112] Paracoccidioidomycosis is the most common systemic mycosis in Latin America and is endemic in several countries, with the highest incidence in Brazil (155 cases per year), followed by Colombia, Venezuela, Uruguay, and Guatemala.[113,114] Active, untreated disease is usually progressive and fatal.[114] Naturally acquired disease has been described only in humans;[115] however, mice, guinea pigs, rats, hamsters, and rabbits have been shown to be susceptible to experimental infections.[115,116] Natural infection in humans is considered to occur from inhalation of the saprophytic form of *P. brasiliensis* from its putative habitat in the soil.[114] Isolation of *P. brasiliensis* from nature has been very limited[117,118] and such reports have been difficult to repeat or confirm. Endemic areas have been identified by incidence of paracoccidioidomycosis as well as paracoccidioidin skin-testing surveys and were associated with subtropical or tropical forest areas.[114] Although paracoccidioidomycosis is 13 to 70 times more common in men than women,[114] contact with the fungus is essentially the same for the two sexes[119] suggesting that sex hormones may play a role in pathogenesis.[120]

The first experimental infections with *P. brasiliensis* were done using guinea pigs more than 70 years ago.[121] Later the hamster was found to be more susceptible and dissemination to the lungs was reported.[122] Recently, models of pulmonary paracoccidioidomycosis in mice have been developed using i.t. or i.n. instillation of the yeast-form.[123-125] Models of acute fatal disease in young mice and chronic progressive disseminated infection in adult mice[125] have been employed for drug therapy[126,127] and immunological studies.[125,126] These models will be described in detail below.

B. Murine Models of Paracoccidioidomycosis
1. Pulmonary Paracoccidioidomycosis
a. Introduction

Attempts to develop a murine model of pulmonary paracoccidioidomycosis were made some 25 years ago by Mackinnon[128] and more recently by Linares et al.;[129] but success has awaited several new developments. Viability and quantitation of a suitable inoculum has been a serious difficulty for the production of a reproducible and useful model. Another experimental problem in working with *P. brasiliensis* has been its poor plating efficiency on standard mycological media. Resolution of these difficulties in the past few years[131,132] has allowed the development of a reproducible model of acute pulmonary or chronic pulmonary disseminated paracoccidioidomycosis.[125-127]

b. Paracoccidioides brasiliensis (Isolates)

Isolates from patients in Medellin, Colombia, that have been used in our work,[125] have been identified by the name or initials of the patient from whom they were isolated, e.g., Garcia, LA, etc. The yeast-form of *P. brasiliensis* can be readily grown on Sabouraud's dextrose agar slants at 35°C. Modified McVeigh Morton broth,[130] inoculated with growth from Sabouraud's dextrose agar and incubated at 35°C on a gyratory shaker, supports good growth in a chemically defined liquid medium.[130] Some isolates, when grown by this method, are more suitable for inoculum preparation than others. For example, the Garcia isolate, when grown in multiple 10 mℓ plastic tubes containing 3 mℓ of McVeigh Morton broth for 7 days at 35°C on a shaker (125 rpm), yielded 63 to 83% single cells.[136,137] In contrast, other isolates under similar conditions yielded only a small percent of single cells; however, an acceptable inoculum could be prepared by a brief sonication.[125,131] Viability of *P. brasiliensis* cells in an inoculum was measured using the ethidium bromide-fluorescein diacetate stain[131] and CFU were assessed by plating triplicate 1 mℓ volumes on McVeigh Morton agar supplemented with 4% horse serum and 5% culture filtrate from 1- to 2-week-old *P. brasiliensis* McVeigh Morton broth cultures.

In this system, cellular viability was approximately 95%, and the plating efficiency with the Garcia isolate was 85%.[126,127,131] A critical factor in defining a reproducible murine model was the finding that when virulence of *P. brasiliensis* was lost by repeated and prolonged in vitro culturing, it was restored upon animal passage. The desired virulence of the Garcia isolate was obtained by 4 to 8 successive transfers in McVeigh Morton broth. Although virulence of different isolates in mice has been reported,[124-126] this could reflect their history of in vitro culture and not their inherent virulence.

For pulmonary infection, mice were anesthetized, as described for infection with *B. dermatitidis* above, and 0.03 m*l* of inoculum instilled intranasally. When a high dose was used, the inoculum was split into two 0.03 m*l* volumes administered within 1 hr. Although delivery of CFU to the lungs was relatively low (10 to 20%), reproducible infections and recovery of CFU was obtained with time.[126,127]

c. Mouse Strains and Age

The advantage of a murine model of pulmonary paracoccidioidomycosis is the availability of inbred strains. To date this has not been exploited and studies with pulmonary paracoccidioidomycosis has been limited to BALB/c,[124-127] NCI,[125] or CD1 Swiss white mice.[128,129] In some studies both male and female mice were infected[123,124] with yeast-phase *P. brasiliensis* and, as we have also observed, there was no noticeable difference in pathogenesis (Castañeda et al.[133]). The maturity of mice was shown to be a critical factor in resistance to acute pulmonary disease. Mice, less than or equal to 15 g (3 to 4 weeks old), had 100% mortality in 10 days with a dose that produced 0% mortality in adult, 20 to 25 g (8- to 12-week-old), mice at 35 days.[125]

d. Histopathology

Histopathology in murine pulmonary paracoccidioidomycosis was described by Mackinnon in 1959,[128] noting bronchoalveolitis 4 days postinfection and granulomatous reactions without caseation after 160 days of chronic infection. These observations have been confirmed and extended in more recent studies by Defaveri et al.[123] and Brummer et al.[125] The histopathology of pulmonary paracoccidioidomycosis in athymic (nu/nu) BALB/c mice has been described by Robledo et al.[124] They noted a lack of granulomatous reactions and giant cell formation.

e. Applications

Several different types of studies have been made possible with the development of an acute fatal or chronic pulmonary disseminated model of paracoccidioidomycosis. Both models have already been employed in studies evaluating the therapeutic efficacy of current drugs (ketoconazole and amphotericin B) vs. new drugs.[126,127] Preliminary immunological studies have also been reported[123,125,126] during the course of chronic pulmonary disseminated paracoccidioidomycosis; however, more systemic studies are possible and are in progress.[133] Development of an immunization protocol which will render mice resistant to pulmonary challenge has been undertaken with the view of eventually protecting humans at risk by vaccination.[134]

2. Intraperitoneal, Intravenous, and Subcutaneous Models of Infection
a. Introduction

Infection of mice with *P. brasiliensis* by the intravenous or peritoneal routes served certain purposes, especially in earlier studies;[128] however, as Linares and Friedman[115] point out, it is not likely that *P. brasiliensis* would gain entry into the body by such routes, and it is difficult to draw conclusions about pathogenesis of paracoccidioidomycosis from observation of experimental infections established by such routes in mice.

b. Intravenous Route

This route of infection with yeast-form *P. brasiliensis* was very effective and produced a high rate of infection[115,128] and mortality (16 of 17 mice).[128] Linares and Friedman[129] found intravenous infection of mice useful in demonstrating the effects of corticosteroid treatment on resistance to this infection and also the virulence of several

P. brasiliensis isolates. Others[135] have used this route of infection to compare granuloma formation in athymic (nu/nu) mice to that of heterozygous (nu/+) littermates.

c. Intraperitoneal and Subcutaneous Routes

A limited number of studies have used the i.p. route of infection with *P. brasiliensis*. Both Mackinnon[128] and Linares et al.[115] compared the i.p. route to other routes of infection for infectivity and mortality. In both studies, infection was established in all the mice; however, mortality was less (13 of 41) than by the i.v. route (17 of 18).[128] A large dose of *P. brasiliensis* (20×10^6) i.p. was used successfully by Kerr et al.[136] to show the immunosuppressive effect of irradiation or cyclophosphamide treatment on severity of infection and dissemination to liver and lungs. On the other hand, Mackinnon found that inoculation by the subcutaneous route was not an effective method of infection and no lesions were found in 10 of 10 mice 4 months later.[128] However, for this very reason, the subcutaneous route may have potential as a method of immunization because in other systems resolution of such types of infection resulted in profound resistance to challenge by the pulmonary[137] or i.v. routes.[138]

C. Other Animal Models of Paracoccidioidomycosis

1. Introduction

Next to the mouse, guinea pigs and hamsters have been used most often as experimental animals to study infection with *P. brasiliensis*. In a series of experiments with five different species of experimental animals, Pollak et al.[116] found rabbits the least susceptible to infection, whereas hamsters and guinea pigs had dissemination from subcutaneous infection sites. In this study, it was found that rats and mice were susceptible by the intraperitoneal route. Although the guinea pig was used first[112] and in early studies,[139] the hamster, because of its greater susceptibility to *P. brasiliensis*,[122] has emerged as a more useful experimental animal.

2. Hamsters

Because of its greater susceptibility to *P. brasiliensis*, the Syrian hamster *(Mesocrietus auratus)* has been used to study the histopathology and dissemination of *P. brasiliensis* from testicular inoculation sites to lymph nodes, liver, spleen, and lungs. An inoculum of 200,000 yeast cells, intratesticularly, resulted in dissemination in 100% of the hamsters after the first week.[140,141] This feature produces a model of chronic progressive disease allowing for measurements of humoral and cellular immune responses[140,141] in a disease state somewhat analogous to the human disease. To date, infection of hamsters by the pulmonary route, and putative natural route of infection, has not been reported. Potentially, this should be a very productive area of investigation.

D. Infection with Mycelial Form

1. Introduction

The majority of studies with *P. brasiliensis* in animals has been done with the yeast-form and this has been so for several reasons. Some of the most obvious are (1) quantitation of inocula and (2) greater infectivity of the yeast-form by virtue of by-passing the mycelial to yeast conversion.[114] However, several basic questions about the putative natural infective process require studies using the saprophytic form of *P. brasiliensis*.

2. Mice

Infecting mice by the pulmonary route with elements (arthroconidia, chlamydoconidia, and arthroaleurioconidia) of the saphrophytic form (120,000 CFU) was successful in cortisone-treated mice.[143] In inoculated mice, 38% developed pulmonary infections, and 14% had dissemination to liver and spleen. This study established the fact that the mycelial phase of *P. brasiliensis* can give rise to pulmonary paracoccidioidomycosis in mice. Full development of this promising model of natural *P. brasiliensis* pulmonary infection awaits advances in inocula preparation and quantitation.

3. Guinea Pigs

Conversion of mycelial to yeast-forms of *P. brasiliensis* in culture at 37°C has been described many times; however, there are only a few systematic studies of mycelial to yeast (M → Y) conversion in tissue;[143,144] Carbonell and Rodriguez[144] made histological studies of the mycelial to yeast conversion in guinea pigs inoculated intratesticularly or subcutaneously with an unquantitated chlamydoconidia containing mycelial suspension. They found that few of the hyphae converted to the yeast form. Acute inflammation with severe polymorphonuclear-granulocyte infiltration at inoculation sites was observed and may have been a factor in poor M → Y conversion. Nevertheless, typical yeast were found in 6 of 12 testicles examined. These preliminary studies should provide a starting point for more quantitative experiments which should include the lung as a study site as well as include females in experiments to consider a reported influence of estrogens on M → Y conversions in vitro.[145]

REFERENCES

1. Gilchrist, T. C. A., A case of blastomycosis dermatitidis in man, *Johns Hopkins Hosp. Rep.*, 1, 269, 1896.
2. Baker, R. D., Tissue reactions in human blastomycosis: an analysis of tissue from twenty three cases, *Am. J. Pathol.*, 18, 479, 1942.
3. Sarosi, G. A. and Davies, S. R., Blastomycosis, *Am. Rev. Respir. Dis.*, 120, 911, 1979.
4. Tenenbaum, M. J. and Kerkering, T. M., Blastomycosis, *CRC Crit. Rev. Microbiol.*, 1982, 139.
5. Furculow, M. L., Smith, C. D., and Turner, C., Supportive evidence by field testing and laboratory experiments for a new hypothesis of the ecology and pathogenicity of canine blastomycosis, *Sabouraudia*, 12, 22, 1974.
6. Sarosi, G. A., Eckman, M. R., Davies, S. F., and Laskey, W., Canine blastomycosis as a harbinger of human disease, *Ann. Intern. Med.*, 91, 733, 1979.
7. Denton, J. F. and DiSalvo, A. F., Respiratory infection of laboratory animals with conidia of *Blastomyces dermatitidis*, *Mycopathol. Mycol. Appl.*, 36, 129, 1968.
8. Harvey, R. P., Schmid, E. S., Carrington, C. C., and Stevens, D. A., Mouse model of pulmonary blastomycosis: utility, simplicity, and quantitative parameters, *Am. Rev. Respir. Dis.*, 117, 695, 1978.
9. Witorsch, P. and Utz, J. P., North American blastomycosis: a study of 40 patients, *Medicine*, 47, 169, 1968.
10. DiSalvo, A. F. and Denton, J. F., Lipid content of four strains of *Blastomyces dermatitidis* of different mouse virulence, *J. Bacteriol.*, 85, 927, 1963.
11. Brummer, E., Morozumi, P. A., Philpott, D. E., and Stevens, D. A., Virulence of fungi: correlation of virulence of *Blastomyces dermatitidis in vivo* with escape from macrophage inhibition of replication *in vitro*, *Infect. Immun.*, 32, 864, 1981.
12. Schwarz, J. and Baum, G. L., Blastomycosis, *Am. J. Clin. Pathol.*, 11, 999, 1951.
13. Denton, J. F., McDonough, E. S., Ajello, L., and Ausherman, R. J., Isolation of *Blastomyces dermatitidis* from soil, *Science*, 133, 1126, 1961.
14. Denton, J. F. and DiSalvo, A. F., Isolation of *Blastomyces dermatitidis* from natural sites in Augusta, Georgia, *Am. J. Trop. Med. Hyg.*, 13, 716, 1964.
15. McDonough, E. S., Dubato, J. J., and Wisniewski, T. R., Soil *Streptomyces* and bacteria related to lysis of *Blastomyces dermatitidis*, *Sabouraudia*, 11, 244, 1974.
16. Dixon, D. M., Shadomy, H. J., and Shadomy, S., *In vitro* growth and sporulation of *Blastomyces dermatitidis* on woody plant material, *Mycologia*, 59, 1193, 1977.
17. Gilchrist, T. C. and Stokes, W. R., Case of pseudo-lupus vulgaris caused by blastomycosis, *J. Exp. Med.*, 3, 53, 1898.

18. Morozumi, P. A., Halpern, J. W., and Stevens, D. A., Susceptibility differences of inbred strains of mice to blastomycosis, *Infect. Immun.*, 32, 160, 1981.

19. Brass, C. and Stevens, D. A., Maturity as a critical determinant of resistance to fungal infections: studies in murine blastomycosis, *Infect. Immun.*, 36, 387, 1982.

20. Morozumi, P. A., Brummer, E., and Stevens, D. A., Protection against pulmonary blastomycosis: correlation with cellular and humoral immunity in mice after subcutaneous nonlethal infection, *Infect. Immun.*, 137, 670, 1982.

21. Brummer, E., Morozumi, P. A., Vo, P. T., and Stevens, D. A., Protection against pulmonary blastomycosis: adoptive transfer with T lymphocytes, but not serum, from resistant mice, *Cell. Immunol.*, 73, 349, 1982.

22. Harvey, R. P., Isenberg, R. A., and Stevens, D. A., Molecular modifications of imidazole compounds: studies of activity and synergy *in vitro* and of pharmacology and therapy of blastomycosis in a mouse model, *Rev. Inf. Dis.*, 2, 559, 1980.

23. Morozumi, P. A., Brummer, E., and Stevens, D. A., Immunostimulation with muramyl dipeptide and its desmethyl analogue: studies of non-specific resistance to pulmonary blastomycosis in inbred mouse strains, *Mycopathologia*, 81, 35, 1983.

24. Hoeprich, P. D. and Finn, P. D., Obfuscation of the activity of antifungal antibiotics by culture media, *J. Infect. Dis.*, 126, 353, 1972.

25. Brass, C., Volkmann, C. M., Klein, H. P., Halde, C. J., Archibald, R. W., and Stevens, D. A., Pathogen factors and host factors in murine pulmonary blastomycosis, *Mycopathologia*, 78, 129, 1982.

26. Bowen, J. T. and Wolbach, S. B., A case of blastomycosis: the results of culture and inoculation experiments, *J. Med. Res.*, 15, 167, 1906.

27. Baker, R. D., Experimental blastomycosis in mice, *Am. J. Path.*, 18, 463, 1942.

28. Spencer, H. D. and Cozad, G. C., Role of delayed hypersensitive in blastomycosis of mice, *Infect. Immun.*, 7, 329, 1973.

29. Brummer, E., Morozumi, P. A., and Stevens, D. A., Macrophages and fungi: *in vitro* effects of method of macrophage induction, activation by different stimuli and soluble factors on *Blastomyces*, *RES J. Reticuloendothel. Soc.*, 28, 507, 1980.

30. Brummer, E. and Stevens, D. A., Enhancing effect of murine polymorphonuclear neutrophils (PMN) on the multiplication of *Blastomyces dermatitidis in vitro* and *in vivo*, *Clin. Exp. Immunol.*, 54, 587, 1983.

31. Brummer, E. and Stevens, D. A., Activation of murine polymorphonuclear neutrophils for fungicidal activity with supernatants from antigen-stimulated immune spleen cell cultures, *Infect. Immun.*, 45, 447, 1984.

32. Cozad, G. C. and Chang, C., Cell-mediated immunoprotection in blastomycosis, *Infect. Immun.*, 28, 398, 1980.

33. Morozumi, P. A., Brummer, E., and Stevens, D. A., Strain differences in resistance to infection reversed by route of challenge: studies in blastomycosis, *Infect. Immun.*, 34, 623, 1981.

34. Baker, R. D., Comparison of infection of mice by mycelial and yeast forms of *Blastomyces dermatitidis*, *J. Infect. Dis.*, 63, 324, 1938.

35. Landay, M. E., Mitten, J., and Miller, J., Disseminated blastomycosis in hamsters. II. Effect of sex on susceptibility, *Mycopath. Mycologia*, 42, 73, 1970.

36. Smith, C. D., Brandsberg, J. W., Selby, L. A., and Menges, R. W., A comparison of the relative susceptibilities of laboratory animals to infection with the mycelial phase of *Blastomyces dermatitidis*, *Sabouraudia*, 5, 126, 1966.

37. Smith, C. D. and Furcolow, M. L., The demonstration of growth stimulating substances for both *Histoplasma capsulatum* and *Blastomyces dermatitidis* in infusion of starling (*Sturnis vulgaris*) manure, *Mycopathologia*, 22, 73, 1963.

38. Smith, C. D., Yeast extract contains growth stimulating substances similar to those in starling manure for *Histoplasma capsulatum* and *Blastomyces dermatitidis*, *Mycopathologia*, 22, 99, 1964.

39. Salfelder, K., Experimental cutaneous North American blastomycosis in hamsters, *J. Invest. Derm.*, 45, 409, 1965.

40. Conant, N. F., Smith, D. T., Baker, R. D., Calloway, J. L., and Martin, D. S., *Manual of Clinical Mycology*, 2nd ed., W. B. Saunders, Philadelphia, 1954.

41. Posadas, A., Un nuevo caso de micosis fungordea con psorospermias, *An. Circ. Med. Argent.*, 15, 585, 1892.

42. Wernicke, R., Ueber einen Protozoenbefund bei Mycosis Fungoides, *Zentrabl. Bakteriol.*, 12, 859, 1892.

43. Rixford, E., A case of protozoic dermatitis, *Occident. Med. Times*, 8, 704, 1894.

44. Rixford, E. and Gilchrist, T. C., Two cases of protozoan (coccidioidal) infection of the skin and other organs, *Johns Hopkins Hosp. Rep.*, 1, 209, 1896.

45. Ophüls, W. and Moffitt, H. C., A new pathogenic mould (formerly described as a protozoan: *Coccidioides immitis* pyogenes), *Philadel. Med. J.*, 5, 1471, 1900.
46. Fiese, M. J., *Coccidioidomycosis*, Charles C Thomas, Springfield, Ill., 1958.
47. Stevens, D. A., Ed., *Coccidioidomycosis*, A Text, Plenum Medical, New York, 1980.
48. Smith, C. E., Reminiscences of the flying chlamydospore and its allies, in *Coccidioidomycosis*, Ajello, L., Ed., University of Arizona Press, Tucson, 1967.
49. Drutz, D. J. and Cantanzaro, A., Coccidioidomycosis, *Am. Rev. Resp. Dis.*, 117, 559 and 727, 1978.
50. Drutz, D. J. and Huppert, M., Coccidioidomycosis: factors affecting the host-parasite interaction, *J. Infect. Dis.*, 147, 372, 1983.
51. Huppert, M., Sun, S. H., and Gross, A. J., Evaluation of an experimental animal model for testing antifungal substances, *Antimicrob. Agents Chemother.*, 1, 367, 1972.
52. Cox, R. A., Pavey, E. F., and Mead, C. G., Course of coccidioidomycosis in intratracheally infected guinea pigs, *Infect. Immun.*, 31, 679, 1981.
53. Levine, H. B., Cobb, J. M., and Smith, C. E., Immunity to coccidioidomycosis induced in mice by purified spherules, arthrospore, and mycelial vaccines, *Trans. N.Y. Acad. Sci.*, 22, 436, 1960.
54. Levine, H. B., Purification of the spherule-endospore phase of *Coccidioides immitis*, *Sabouraudia*, 1, 112, 1961.
55. Levine, H. B., A direct comparison of oral treatments with Bay-n-7133, Bay-1-9139 and ketoconazole in experimental murine coccidioidomycosis, *Sabouraudia*, 22, 34, 1984.
56. Clemons, K. V., Leathers, C. R., and Lee, K. W., Systemic *Coccidioides immitis* infection in nude and beige mice, *Infect. Immun.*, 47, 814, 1985.
57. Pappagianis, D., Levine, H. B., Smith, C. E., Berman, R. J., and Kobayashi, G. S., Immunization of mice with viable *Coccidioides immitis*, *J. Immunol.*, 86, 28, 1961.
58. Huppert, M., Sun, S. H., Gleason-Jordan, I., and Vukovich, K. P., Lung weight parallels disease severity in experimental coccidioidomycosis, *Infect. Immun.*, 14, 1356, 1976.
59. Kong, Y. M., Levine, H. B., Madin, S. H., and Smith, C. E., Fungal multiplication and histopathological changes in vaccinated mice infected with *Coccidioides immitis.*, *J. Immunol.*, 92, 779, 1964.
60. Savage, D. C. and Madin, S. H., Cellular responses in lungs of immunized mice to intranasal infection with *Coccidioides immitis*, *Sabouraudia*, 6, 94, 1968.
61. Sinski, J. T. and Soto, P. J., Onset of coccidioidomycosis in mouse lung after intravenous injection, *Mycopathologia*, 30, 41, 1966.
62. Kirkland, T. N. and Fierer, J., Cyclosporin A inhibits *Coccidioides immitis in vitro* and *in vivo*, *Antimicrob. Agents Chemother.*, 24, 921, 1983.
63. Pappagianis, D., Hector, R., Levine, H. B., and Collins, M. S., Immunization of mice against coccidioidomycosis with a subcellular vaccine, *Infect. Immun.*, 25, 440, 1979.
64. Kong, Y. M., Levine, H. B., and Smith, C. E., Immunogenic properties of nondisrupted and disrupted spherules of *Coccidioides immitis* in mice, *Sabouraudia*, 2, 131, 1963.
65. Kirkland, T. N. and Fierer, J., Inbred mouse strains differ in resistance to lethal *Coccidioides immitis* infection, *Infect. Immun.*, 40, 912, 1983.
66. Scalarone, G. M., and Huntington, R. W., Circling syndrome and inner ear disease in mice infected intraperitoneally or intravenously with *Coccidioides immitis* spherule-endospore cultures, *Mycopathologia*, 86, 75, 1983.
67. Lecara, G., Cox, R. A., and Simpson, R. B., *Coccidioides immitis* vaccine: potential of an alkali-soluble, water-soluble cell wall antigen, *Infect. Immun.*, 39, 473, 1983.
68. Beaman, L., Pappagianis, D., and Benjamini, E., Mechanisms of resistance to infection with *Coccidioides immitis* in mice, *Infect. Immun.*, 23, 681, 1979.
69. Beaman, L., Pappagianis, D., and Benjamini, E., Significance of T cells in resistance to experimental murine coccidioidomycosis, *Infect. Immun.*, 17, 580, 1977.
70. Morozumi, P. S., Brummer, E., and Stevens, D. A., Strain differences in resistance to infection reversed by route of challenge: studies in blastomycosis, *Infect. Immun.*, 34, 623, 1982.
71. Kong, Y. M. and Levine, H. B., Experimentally induced immunity in the mycoses, *Bacteriol. Rev.*, 31, 35, 1967.
72. Cox, R. A., Mead, C. G., and Pavey, E. F., Comparisons of mycelia- and spherule-derived antigens in cellular immune assays of *Coccidioides immitis*-infected guinea pigs, *Infect. Immun.*, 31, 687, 1981.
73. Levine, H. B., Miller, R. L., and Smith, C. E., Influence of vaccination on respiratory coccidioidal disease in cynomologus monkeys, *J. Immunol.*, 89, 242, 1962.
74. Lowe, E. P., Sinski, J. T., Huppert, M., and Ray, J. G., Coccidioidin skin tests and serologic reactions in immunized and infected monkeys, in *Coccidioidomycosis*, Ajello, L., Ed., University of Tucson Press, Tucson, 1967, 171.
75. Converse, J. L., Lowe, E. P., Castleberry, M. W., Blundell, G. P., and Besemer, A. R., Pathogenesis of *Coccidioides immitis* in monkeys, *J. Bacteriol.*, 83, 871, 1962.

76. Converse, J. L., Pakes, S. P., Snyder, E. M., and Castleberry, M. W., Experimental primary cutaneous coccidioidomycosis in the monkey, *J. Bacteriol.*, 87, 81, 1964.

77. Cronkite, A. E. and Lack, A. R., Primary pulmonary coccidioidomycosis: experimental infection with *Coccidioides immitis, J. Exp. Med.*, 72, 167, 1940.

78. Ibrahim, A. B. and Pappagianis, D., Experimental induction of anergy to coccidioidin by antigens of *Coccidioides immitis, Infect. Immun.*, 7, 786, 1973.

79. Sinski, J. T. and Dalldorf, F. G., Coccidioidin sensitivity in guinea pigs immunized with killed arthrospores, in *Coccidioidomycosis*, Ajello, L., Ed., University of Tucson Press, Tucson, 1967, 175.

80. Likholetov, S. M., Delayed hypersensitivity and transfer factor in experimental coccidioidomycosis, *Sabouraudia*, 17, 251, 1979.

81. Darling, S. T., A protozoan general infection producing pseudotubercles in lungs and focal necrosis in the liver, spleen and lymph nodes, *J. Am. Med. Assoc.*, 46, 1283, 1906.

82. deMonbreun, W. A., The cultivation and cultural characteristics of Darling's *Histoplasma capsulatum, Am. J. Trop. Med.*, 14, 93, 1934.

83. Emmons, C. W., Isolation of *Histoplasma capsulatum* from soil, *Public Health Rep.*, 64, 892, 1949.

84. Zeidlberg, L. D. and Ajello, L., Isolation of *Histoplasma capsulatum* from soil, *Am. J. Public Health*, 42, 930, 1952.

85. Ajello, L. and Manson-Bahr, P. E., Amboni caves, Tanganyika: a new epidemic area for *Histoplasma capsulatum, Am. J. Trop. Med. Hyg.*, 9, 633, 1960.

86. Kessel, B., Distribution and migration of the European starling in North America, *Condor*, 55, 49, 1953.

87. Furcolow, M. L., Recent studies on the epidemiology of histoplasmosis, *Ann. N.Y. Acad. Sci.*, 72, 127, 1958.

88. DiSalvo, A. F. and Johnson, W. M., Histoplasmosis in South Carolina: support for the microfocus concept, *Am. J. Epidemiol.*, 109, 480, 1979.

89. Burk, R. L. and Jones, B. D., Disseminated histoplasmosis with osseous involvement in a dog, *J. Am. Vet. Med. Assoc.*, 172, 1416, 1978.

90. Hall, A. D., An equine abortion due to histoplasmosis, *Vet. Med. Small Animal Clinic*, 74, 200, 1979.

91. DiSalvo, A. F. and Ajello, L., Isolation of *Histoplasma capsulatum* from Arizona bats, *Am. J. Epidemiol.*, 89, 606, 1969.

92. Schwartz, J. and Baum, G. L., Successful infection of pigeons and chickens with *Histoplasma capsulatum, Mycopathologia*, 8, 189, 1957.

93. Rowley, D. A. and Huber, M., Pathogenesis of experimental histoplasmosis in mice. I. Measurement of infection dosages of the yeast phase of *Histoplasma* capsulatum, *J. Infect. Dis.*, 96, 174, 1955.

94. Procknow, J. J., Page, M. I., Clayton, G., and Loosli, M. D., Early pathogenesis of experimental histoplasmosis, *Arch. Pathol.*, 69, 413, 1960.

95. Baughman, R. P., Hendricks, D., and Bullock, W. E., Sequential analysis of cellular immune responses in the lung during *Histoplasma capsulatum* infection, *Clin. Res.*, 33, 425A, 1984.

96. Nickerson, D. A. and Fairclough, P., Immune responsiveness following intratracheal inoculation with *Histoplasma capsulatum* yeast cells, *Clin. Exp. Immunol.*, 56, 337, 1984.

97. Tewari, R. P., Sharma, D., Solotarovsky, M., Lafemina, R., and Balint, J., Adoptive transfer of immunity from mice immunized with ribosomes of live yeast cells of *Histoplasma capsulatum, Infect. Immun.*, 15, 789, 1977.

98. Sun, S. H. and Huppert, M., A cytological study of morphogenesis in *C. immitis, Sabouraudia*, 14, 185, 1976.

99. Berlinger, M. D. and Reca, M., Vital staining of *Histoplasma capsulatum* with Janus green B, *Sabouraudia*, 5, 26, 1966.

100. Burt, W. R., Underwood, A. L., and Appleton, G. L., Hydroxamic acid from *Histoplasma capsulatum* that displays growth factor activity, *Appl. Environ. Microbiol.*, 42, 560, 1981.

101. Schlitzer, R. L., Chandler, F. W., and Larsh, H. S., Primary acute histoplasmosis in guinea pigs exposed to aerosolized *Histoplasma capsulatum, Infect. Immun.*, 33, 575, 1981.

102. Howard, D. H., Otto, V., and Gupta, R. K., Lymphocyte-mediated cellular immunity in histoplasmosis, *Infect. Immun.*, 4, 605, 1971.

103. Burt, W. R. and Smith, R. A., Studies on experimental murine histoplasmosis: host protection and cellular immunity, *Can. J. Microbiol.*, 29, 102, 1982.

104. Tewari, R. P., Sharma, D. K., and Mathur, A., Significance of T-lymphocytes in immunity elicited by immunization with ribosomes or live yeast cells of *Histoplasma capsulatum, J. Inf. Dis.*, 138, 605, 1978.

105. Howard, D. H. and Otto, V., Experiments of lymphocyte-mediated cellular immunity in murine histoplasmosis, *Infect. Immun.*, 16, 226, 1977.

106. Artz, R. P. and Bullock, W. E., Immunoregulatory responses in experimental disseminated histoplasmosis: depression of T-cell dependent and T-cell responses by activation of splenic suppressor cells, *Infect. Immun.*, 23, 893, 1979.

107. Williams, D. M., Graybill, J. R., and Drutz, D. J., *Histoplasma capsulatum* infection in nude mice, *Infect. Immun.*, 21, 973, 1978.

108. Miyaji, M., Chandler, F. W., and Ajello, L., Experimental histoplasmosis capsulation in athymic nude mice, *Mycopathologia*, 75, 139, 1981.

109. Turner, C., Furcolow, M. L., and Smith, C. D., Experimental histoplasmosis and blastomycosis in young pups, *Sabouraudia*, 12, 188, 1974.

110. Salvin, S. F., Cultural and serological studies on nonfatal histoplasmosis in mice, hamsters and guinea pigs, *J. Infect. Dis.*, 94, 222, 1954.

111. Daniels, L. S. and Berlinger, M. D., Varying virulence in rabbits infected with different filamentous types of *Histoplasma capsulatum.*, *J. Bacteriol.*, 96, 1535, 1968.

112. Lutz, A., Uma mucose psuedo-coccidicia localisada no boca e observada no Brazil: contirbuiaco ao conhecimento das hypho-blastomycoses americanas, *Bras. Med.*, 22, 121, 1908.

113. Restrepo, A., Paracoccidioidomycosis, *Acta Med. Col.*, 3, 33, 1978.

114. Restrepo, A., Greer, D. L., and Vasconcellos, M., Paracoccidioidomycosis: a review, *Rev. Med. Vet. Mycol.*, 8, 97, 1973.

115. Linares, L. and Friedman, L., Pathogenesis of paracoccidioidomycosis in experimental animals, Proc. First Pan Am. Symp., PAHO and WHO Scientific Publication No. 254, Washington, D.C., 1972, 287.

116. Pollak, L. and Angulo-Ortega, A., Pathogenesis of paracoccidioidomycosis, Proc. First Pan Am. Symp., PAHO and WHO Scientific Publication No. 254, Washington, D.C., 1972, 293.

117. Albornos, M., Isolation of *Paracoccidioides brasiliensis* from rural soil in Venezuela, *Sabouraudia*, 9, 248, 1971.

118. Grose, E. and Tamsitt, J. R., *Paracoccidioides brasiliensis* recovered from the intestinal tract of the fructivorous bat, *A. Lituratus, Sabouraudia*, 4, 124, 1965.

119. Restrepo, A., Robledo, M., Ospina, S., Restrepo, M., and Correa, A., Distribution of paracoccidioidin sensitivity in Colombia, *Am. J. Trop. Med.*, 17, 25, 1968.

120. Loose, D. S., Stover, E. P., Restrepo, A., Stevens, D. A., and Feldman, D., Estradiol binds to a receptor-like cytosol binding protein and initiates a biological response in *Paracoccidioides brasiliensis, Proc. Nat. Acad. Sci. U.S.A.*, 80, 7659, 1983.

121. Percira, M. and Vianna, G., A proposito de um caso de blastomycose (Pyohemia blastomycotica), *Arq. Brasil Med.*, 1, 63, 1911.

122. Guimeraes, F., Infeccao do hamster (*Cricetus auratus Waterhouse*) pelo agente da micose de Lutz (blastomicose sul-americana), *Hospital (Rio)* 40, 515, 1951.

123. Defaveri, J., Rezkallah-Iwasso, T., and Franco, M. F., Experimental pulmonary paracoccidioidomycosis in mice: morphology and correlation of lesions with humoral and cellular immune response, *Mycopathologia*, 77, 3, 1982.

124. Robledo, M. A., Graybill, J. R., Ahrens, J., Restrepo, A., Drutz, D., and Robledo, M., Host defenses against experimental paracoccidioidomycosis, *Am. Rev. Respir. Dis.*, 125, 563, 1982.

125. Brummer, E., Restrepo, A., Stevens, D. A., Azzi, R., Gomez, A., Hoyos, G., McEwen, J., Cano, L., and deBedout, C., Murine model of paracoccidioidomycosis. Production of fatal acute pulmonary or chronic pulmonary and disseminated disease: immunological and pathological observations, *J. Exp. Path.*, 1, 241, 1984.

126. Hoyos, G., McEwen, J., Brummer, E., Castañeda, E., Restrepo, A., and Stevens, D. A., Chronic murine paracoccidioidomycosis: effect of ketoconazole on clearance of *Paracoccidioides brasiliensis* and immune response, *Sabouraudia*, 22, 419, 1984.

127. Lefler, E., Brummer, E., McEwen, J., Hoyos, G., Restrepo, A., and Stevens, D. A., Study of current and new drugs in a murine model of acute paracoccidioidomycosis, *Am. J. Trop. Med. Hyg.*, 34, 134, 1985.

128. Mackinnon, J. E., Blastomicosis Sudamericana experimental evolution por via pulmonar, *Ann. Fac. Med. (Montevideo)*, 44, 355, 1959.

129. Linares, L. I. and Friedman, L., Experimental paracoccidioidomycosis in mice, *Infect. Immun.*, 5, 681, 1972.

130. Restrepo, A. and Jiminez, B. E., Growth of *P. brasiliensis* yeast phase in a chemically defined culture medium, *J. Clin. Microbiol.*, 12, 279, 1980.

131. Restrepo, A., Gano, L., deBedout, C., Brummer, E., and Stevens, D. A., Comparison of various techniques to determine the viability of *P. brasiliensis* yeast cells, *J. Clin. Microbiol.*, 16, 204, 1982.

132. Castañeda, E., Brummer, E., and Stevens, D. A., Development of a culture medium for *Paracoccidioides brasiliensis* with high plating efficiency, manuscript in preparation.

133. Castañeda, E., Brummer, E., Pappagianis, D., and Stevens, D. A., Progression, chronicity, immune responses and mortality in murine pulmonary paracoccidioidomycosis, *Abstr. Annu. Meet. Am. Soc. Microbiol.*, F16, 79, 1985.
134. Castañeda, E., Brummer, E., Pappagianis, D., and Stevens, D. A., Resistance to murine pulmonary paracoccidioidomycosis provided by vaccination, *Abstr. Int. Congr. ISHAM,* P3-35, 23, 1985.
135. Miyaji, M. and Nishimura, K., Granuloma formation and killing function of granuloma in congenitally athymic nude mice infected with *Blastomyces dermatitidis* and *Paracoccidioides brasiliensis,* *Mycopathologia*, 82, 129, 1983.
136. Kerr, I. B., daCosta, S. C., and Alencar, A., Experimental paracoccidioidomycosis in immunosuppressed mice, *Immunol. Lett.*, 5, 151, 1982.
137. Morozumi, P. A., Brummer, E., and Stevens, D. A., Protection against pulmonary blastomycosis: correlation with cellular and humoral immunity in mice after subcutaneous nonlethal infection, *Infect. Immun.*, 137, 670, 1982.
138. Tewari, R. P., Sharma, D., Solotorovsky, M., Lafemina, M., and Balint, J., Adoptive transfer of immunity from mice immunized with ribosomes or live yeast cells of *Histoplasma capsulatum, Infect. Immun.*, 15, 739, 1977.
139. DeBrito, T. and Netto, C. F., Disseminated experimental South American blastomycosis of the guinea pig: a pathologic and immunologic study, *Path. Microbiol.*, 26, 29, 1963.
140. DelNegro, G., Lacaz, C., and Fiorillo, A., *Paracoccidioidomycosis: Blastomicose Sub-America,* Pub. Sarvier-Edusp., Sao Paulo, Brazil, 1982, 78.
141. Iabuki, K. and Montenegro, M. R., Experimental paracoccidioidomycosis in the Syrian hamster: morphology, ultrastructure and correlation of lesions with presence of specific antigen and serum levels of antibodies, *Mycopathologia*, 67, 131, 1979.
142. Peracoli, M. T., Mota, N. G., and Montenegro, M. R., Experimental paracoccidioidomycosis in the Syrian hamster: morphology and correlation of lesions with humoral and cell mediated immunity, *Mycopathologia*, 79, 7, 1982.
143. Restrepo, A. and deGuzman, E. G., Paracoccidioidomycosis experimental del ration inducida por via aerogena, *Sabouraudia*, 14, 299, 1976.
144. Carbonell, L. and Rodriguez, J., Transformation of mycelial and yeast forms of *Paracoccidioides brasiliensis* in cultures and in experimental inoculations, *J. Bacteriol.*, 90, 504, 1965.
145. Restrepo, A., Salazar, M. E., Cano, L. E., Stover, E. P., Feldman, D., and Stevens, D. A., Estrogens inhibit mycelium-to-yeast transformation in the fungus *Paracoccidioides brasiliensis:* implications for resistance of females to paracoccidioidomycosis, *Infect. Immun.*, 46, 346, 1984.
146. Brummer, E. and Clemons, K. V., unpublished observations.

Chapter 4

OPPORTUNISTIC FUNGAL INFECTIONS

H. Yamaguchi

TABLE OF CONTENTS

I. ASPERGILLOSIS

A. Introduction

Aspergillosis can be defined as damage to tissue caused by several species of *Aspergillus*. Damage may be caused by invasion of tissues, allergic mechanisms, or colonization of cavities, resulting in the diseases varying in severity and clinical course which may also depend upon the organs affected and the host. Apergilli are abundant in the environment. The genus *Aspergillus* now includes more than 130 recognized species,[1] all of which are world-wide in distribution. Their spores are produced in great abundance and are readily disseminated into the air by wind current. Although the spores are frequently inhaled by man, most of human aspergillosis are thought to be caused by members of the *A. fumigatus* group. However, members of other groups, particularly, *A. flavus*, *A. niger* and *A. terreus*, have also been implicated.[1-3]

The lung is the most common site affected. Inhalation of spores can give rise to a number of different clinical forms of aspergillosis. Although an individual with apparently normal defense mechanisms may be rarely infected by a member of the *A. fumigatus* group ("primary invasive form"), invasive aspergillosis mostly occurs in individuals whose resistance is lowered as a result of severe debilitating diseases, especially leukemia and other hematological neoplasma, lymphoma, renal transplants, and chronic granulomatous disease and in individuals receiving corticosteroids, immunosuppressive drugs, and broad-spectrum antibiotics, alone or in combination, that have been most commonly recognized as predisposing agents ("secondary invasive form").[3-9] From clinical observations it has been concluded that neutropenia and high doses of corticosteroids are the two major risk factors for invasive aspergillosis[4,5] and that the combination may act synergistically to break down natural resistance.[5] This form of invasive aspergillosis is mainly implicated by members of the *A. fumigatus* group, with *A. flavus* as the next most important pathogen.[3] Its increasing incidence in recent years raises a serious problem in controls of immunocompromised patients. In patients with this form of aspergillosis, dissemination of the infection to other parts of the body may also occur. The most common sites for the disseminated lesions are the brain, kidney, liver, gastrointestinal tract, and myocardium. In both the pulmonary and the disseminated disease of other organs, the fungus invades blood vessels and causes thrombosis and infarcts.[4,5]

Preexisting or concurrent lung disease is a major factor predisposing a host to *A. fumigatus* colonization. A tuberculous cavity, lung abscess, cavitating infarcts, lung cysts, bronchiectasis, or sarcoidosis all predispose the patient to *Aspergillus* colonization.[10,11] By the time the condition is diagnosable, the fungus has usually grown to form a compact mass of mycelium termed "fungus ball" or "aspergilloma". Members of the *A. niger* and *A. fumigatus* groups are most often involved in this form of aspergillosis. In most cases in the noncompromised host, the fungus does not invade surrounding tissue and spread to other parts of the body. In contrast, if an aspergilloma should occur in an immunocompromised host, or an otherwise healthy patient with aspergilloma should receive immunosuppression, a difficult and hazardous clinical problem is posed; in this situation, an aspergilloma may rapidly increase in size, or may become frankly invasive, producing necrotizing *Aspergillus* pneumonitis with or without hematogenous dissemination. This seems more likely to occur if the aspergilloma is not enclosed in a thick capsule of fibrous tissue.

Allergic bronchopulmonary aspergillosis (ABA) is another major form of pulmonary aspergillosis. This allergic disease is being recognized with increasing frequency.[12,13] It occurs in previously sensitized persons who are exposed to the spores of an *Aspergillus* sp., especially members of the *A. fumigatus* group being the usual etiologic agents. In ABA, febrile episodes occur in association with transient lung infiltrates recurring in multiple sites. This syndrome is associated with pulmonary consolidation and peripheral blood eosinophilia.

In addition to the various pulmonic forms of aspergillosis, specific clinical types involving other organs are also recognized. Such disorders caused by members of the *A. fumigatus* and *A. flavus* groups include otomycosis, keratomycosis (see Section IV of this chapter) and other ocular mycosis and, rarely, endocarditis.

Aspergillosis, especially pulmonary forms, have been recorded in a wide variety of domestic and wild animals, as well as birds.[14] Birds, in particular, are susceptible to infection by aspergilli.[14] Much of our knowledge of the disease was acquired from reports published prior to 1900. Since the turn of the century, aspergillosis has been mainly associated with young poultry;[15] however, it has been reported in the past few decades in many mammalian species, such as calves,[16,17] horses,[18] dogs, cats,[19] sheep,[20] and deer.[21] The *Aspergillus* species may also infect the placenta of cattle, causing abortion.[14] Natural infection with *Aspergillus* species has been reported only with rabbits among various experimental animals.[22,23]

To gain insight into the pathogenesis of human aspergillosis, a large number of studies have been conducted, using a variety of experimental models of the disease. Fatal aspergillosis involving the kidneys, lung, and other organs have been most frequently produced in mice. Certain murine models have been demonstrated to be effective in evaluating antifungal agents; although less frequently, rabbits and various other species of animals, including rats, monkeys, sheep, ducks and chicks, have also been employed as models for experimental aspergillosis. Mice, rabbits, and other laboratory animals normally show high resistance to invasive aspergillosis that can be lowered by corticosteroids or immunosuppressions.[28-41] Thus administration of large doses of cortisone prior to exposure to *Aspergillus* inoculum is occasionally required in order to secure establishment of persistent and invasive infections in these animals.

B. Murine Models of Aspergillosis

1. Introduction

The most frequently used animal model of aspergillosis is the mouse, infected with spores from varying species of *Aspergillus*, especially *A. fumigatus* and *A. flavus*. Based on the route of inoculation, almost all the murine models thus far developed can be categorized into either one of the following two types: (1) intravenous models and (2) inhalation models. In the latter models, mice are subjected to inhalation of aerosolized dry spores or, less frequently, saline suspensions of spores. This method of exposure probably simulates more closely the mechanism by which human beings acquire fungal infections of the lungs. However, normal mice exposed to the inhalation of *Aspergillus* spores are resistant to lethal infection and, therefore, treatment of the animals with corticosteroids has been shown essential for developing a high incidence of fatal bronchopulmonary aspergillosis in this model.[37,38,42-44] On the other hand, despite the artificial route of infection, the intravenous models constitute the advantages that fatal infections have been successfully produced without steroids and that precise dose-response of *Aspergillus* spores has been established. These make it possible to quantify more directly the virulence of fungal strains and the therapeutic efficacy of antifungal agents avoiding the fluctuating effects of steroids.

2. Intravenous Models

a. Virulence of Strains

When the intravenous route of infection has been used, *A. fumigatus* has been most preferably selected as a challenge organism, although several *Aspergillus* species other than *A. fumigatus*, such as *A. flavus*, *A. flavipes*, *A. nidulans*, and *A. terreus*, have been also employed. With the exception of a very limited number of studies in which aleuriospores from the members of *A. terreus-flavipes* group were used, conidia harvested from mature cultures and suspended in normal saline have been usually used as infecting inoculum. The mortality, survival time, and infection rate of the infected mice appears to depend upon not only the size of inoculum, but also species and strains of *Aspergillus*. Ford and Friedman[35] compared the relative virulence of each strain of 14 species of aspergilli by inoculating normal mice intravenously with graded doses of spores and found that 11 possessed some degrees of virulence, whereas 3 others were avirulent. They also found that only the members of the *A. flavus* group consistently killed mice with doses as low as 10^4 viable spores. These workers were unable to relate the virulence to spore characteristics, such as germination time, size, or shape. Pore and Larsh[45] demonstrated that fatal infection was produced with intravenous inocula of all four members of the *A. terreus-flavipes* group tested, each causing fatal infections with marked neurological signs and histopathological findings of multiple acute aspergillosis of the brain. Morphologically, the most unique feature of these fungi in the lesions was the appearance of aleuriospores, which are the particularly resistant spores exclusively produced by this group of aspergilli.[45,46] Although clinical isolate of *A. fumigatus* has been most often used as a challenge organism, only limited information is available with regard to variances of the virulence among different strains of this speices. Scholer[47] reported that six strains of *A. fumigatus* differed only slightly in virulence and Ford and Friedman[35] found that three strains were comparable with each other in this respect, when tested using the intravenous route of inoculation. Smith[48] demonstrated that 4 of 5 strains of *A. fumigatus* were of closely similar virulence, but that a 5th strain, which grew more slowly in vitro, was less virulent. He has further tested eight strains of *A. fumigatus* under the same experimental conditions and found that the virulence of one strain, which produced unusually large spores, was of greater virulence than the seven other strains.[49] Occurrence of a significant variance in virulence among different clinical isolates have been also reported by Graybill et al.[50]

The symptomatology and histopathology of the fatal disease which follows intravenous inoculation of spores of a virulent *A. fumigatus* strain has been precisely described by several workers.[47,48,50] Although initially large numbers of spores were present in the lung, liver, spleen, kidneys, and brain of the infected animals, infection was progressively lost, first from the lungs, then spleen and liver, and later from brain, so that in animals dying 2 weeks or more after inoculation, infection was almost invariably confined to the kidneys. Rippon et al.[51] reported that a human isolate of *A. terreus* was of greater virulence and grew faster at 37°C in liquid media than soil isolates of *A. terreus.*

In a series of intensive studies of the genetic control of virulence, using a number of genetically-defined strains of *A. nidulans,* Purnell[52-56] has demonstrated the importance of its genetic composition to its virulence to mice. Various mapped point mutations (mostly auxotrophic and euzygotic) were shown to be either neutral in their effect on virulence or were associated with significant reduction in virulence or avirulence.[53,56] On the other hand, a morphologic mutation and diploid strains of this normally haploid fungus were associated with significantly increased virulence.[54,55] These, and the results obtained from his later study showing a different histopathologic response of mice to intravenous inoculation of *A. nidulans* strains with different virulence,[57] led him to the conclusion that the difference in virulence of the strains probably resides in inherent difference in their genetic composition.

b. Effects of Corticosteroids and Other Agents

In an attempt to assess the contribution of natural and acquired immunity to resistance to *Aspergillus* infections, a number of workers have been studying the effects of treatment of the mice with steroids and other immunosuppressive drugs on the course of infection induced by the challenge of *A. fumigatus* spores. Lehman and White[58,59] demonstrated that cortisone pretreatment rendered mice more susceptible to *A. fumigatus* infection and that this was associated with an increased mycelial growth rate in the kidneys, as well as in the liver and heart. These investigators reasoned that a cortisone-insensitive systemic immunity, developing in mice with established kidney infections, would protect the liver and heart against a second conidial challenge given after coritsone treatment. Corbel and Eades[32,60] have reported that resistance of adult mice to *A. fumigatus* infection was markedly lowered by immunosuppressive treatment with prednisone, prednisone plus cyclophosphamide, or antilymphocytic serum, producing severe depression of lymphocyte function and that, in consistence, aging New Zealand Black mice with naturally developing deficiency of cell-mediated immune function were more susceptible to lethal infection with intravenously challenged *A. fumigatus* than similarly aged normal mice of the CBS strain. Schaffner et al.[41] have conducted experiments on various invasive aspergillosis models produced in normal mice, nitrogen mustard-induced neutropenic mice, and athymic nude mice in order to clarify the mechanism of action of corticosteroids toward defenses against *A. fumigatus* infection. The results brought them to the suggestion that natural immunity to *Aspergillus* may be constituted by two sequential lines of defense: the first line of defense formed by macrophages is directed against spores, and the second line of defense is the polymorphonuclear leukocytes (PMN) which protect against the hyphal form of *Aspergillus.* Also, steroids may damage the former directly and not through the influence of T-lymphocytes or other systems modifying macrophage function. The work of Shiraishi[61] also showed a significant role of phagocytosis in the defense of mice against *A. fumigatus* brain infection. Purnell[62] reported that killed *Corynebacterium parvum* vaccine, known as an immunomodulating agent with antitumor activity, did not stimulate but reduced the host resistance to *A. nidulans* infection and facilitated the course of fatal murine aspergillosis.

3. Inhalation Models

Despite the natural route of infection, normal mice experimentally exposed to the inhalation of viable *Aspergillus* spores are resistant to lethal infection.[29,34,37,63] Almost all trials thus far made have failed to establish fatal aspergillosis in mice given *A. fumigatus* spores by the inhalation route without pretreatment of the animals with steroids or other immunosuppressive drugs.[29,30,34,37,40,45,63] Virtually the same results have been obtained when mice were challenged with spores from a virulent strain of *A. flavus* under similar experimental conditions.[29,35,64] Sandhu et al.[44] examined the effect of cortisone on bronchopulmonary aspergillosis in mice receiving inhalation of spores of six *Aspergillus* species, viz., *A. flavus*, *A. fumigatus*, *A. nidulans*, *A. niger*, *A. tamarii*, and *A. terreus*, and subsequently observed a significant enhancement in mortality and tissue invasion due to all the test species in the cortisone-treated animals.

The inhalation murine models of aspergillosis have been most frequently produced in mice infected by inhalation of aerosols of dry viable spores of *A. flavus* or *A. fumigatus*.[37,38,42-44] In these procedures, a culture grown on conventional agar media at 37°C for 2 to 4 days with resultant profuse sporulation is the source of the spores. Mice are placed in the inhalation chamber of Piggot and Emmons[65] or some modified type consisting of a closed bell-jar containing a cylindrical wire mesh.[29] The chamber usually permits simultaneous exposure of ten or more animals to spores aerosolized by blowing air vertically over a culture grown in the chamber.[40,41] With the use of this type of murine model, Sideransky's group of investigators and several others have clearly demonstrated that mice exposed to inhalation of aerosolized *A. flavus* spores became highly susceptible to fatal pulmonary aspergillosis when subjected to treatment with steroids, as well as X-irradiation or immunosuppressive drugs.[29,34,37,63] Similarly, development of increased susceptibility to *A. flavus* pulmonary infection has been observed in alloxan-induced diabetic mice,[34] mice bearing transplantable lymphoid leukemia, or mice fed deficient diets.[35] The effect of corticosteroids which lowers resistance to fatal pulmonary aspergillosis has also been demonstrated in mice infected with *A. fumigatus* spores.[40,41,43,44]

Histopathological examination showed that inhaled *Aspergillus* spores germinate rapidly and profusely into hyphae with subsequent penetration throughout the lungs in cortisone-treated mice, but did not germinate in the lungs of control mice.[29,40,43,44] A number of studies have offered evidence showing that murine alveolar macrophages are capable of preventing germination and killing spores[37,38,40,41,43,66,67] and that corticosteroids may act to stabilize lysosomal membranes within the macrophages ingesting spores, thus interfering with the intracellular destruction of spores, thereby allowing them to germinate.[37,38,67]

The inhalation model of fatal aspergillosis has been also established in mice administered intranasal inoculation of the suspensions of spores from *A. fumigatus* or *A. flavus*.[67,68] Procedures for such a pulmonary challenge were described by Graybill and his coworkers.[50,68] Cortisone-treated mice were anesthetized and abdominal pressure was applied to force expiration. Then a drop of 0.05 m*l* of saline suspension of spores was placed on the nares, and the pressure was released. In the next inspiration, the droplet was inhaled. The successful establishment of pulmonary infection was confirmed by the histopathologic studies which demonstrated formation of abscesses containing hyphae in both lungs a few days after challenge.[68] Using a similar technique of intranasal inoculation, Epstein et al.[67] induced intracranial infection with *A. flavus* in cortisone-treated mice. Histopathologically, widely disseminated infection occurring in the brain parenchyma and leptomeninges was regularly observed.

4. Intraperitoneal Models

Because of the difficulty in intravenous injection or inhalation of germinating spores of *Aspergillus* into mice, Sideransky et al.[36] have used the intraperitoneal route of inoculation to induce fatal infection with this specific form of inoculum. The results of their studies demonstrated that while nongerminating spores of *A. flavus*, administered intraperitoneally into normal mice, induced a low incidence of lethal infection, a similar dose of germinating spores of the same fungus induced a high incidence of fatal disease associated with widely disseminated visceral hyphal aspergillosis. Cortisone-treated mice were again shown to be highly susceptible to fatal infection with nongerminating spores of *A. flavus* inoculated intraperitoneally.[36]

5. Murine Models for Evaluation of Antifungal Agents

A number of antifungal agents which are commercially available or being currently developed have been evaluated in mice experimentally infected with *A. fumigatus*.[50,68,69-75] It was demonstrated that amphotericin B, its D-ornithyl methyl ester, flucytosine, ketoconazole, vibunazole (Bay n 7133), as well as amphotericin B in combination with certain other drugs, were effective in prolonging survival of mice which had been challenged intravenously with lethal doses of *A. fumigatus* spores.[53,68,70,71,73,75] Along with this model, Graybill and his coworkers[53,70] have used an inhalation model produced in cortisone-treated mice to examine the therapeutic efficacy of the two new triazole drugs, vibunazole and itraconazole with positive results. However, these investigators have found the inhalation model inadequate for experimental chemotherapy with amphotericin B or related antibiotics because corticosteroids and polyenes exerted a synergistic toxicity.[68] On the other hand, Schaffner and Frick[69] reported the protective effect of amphotericin B in neutropenic mice which had been challenged intravenously with *A. fumigatus* spores following pretreatment with cortisone plus nitrogen mustard.

C. Rabbit Models

1. Introduction

Next to the mouse, the rabbit has been most frequently used as a model of *Aspergillus* infections. Since Henrici[76] did pioneering work on the biological activity and the etiopathological role of *Aspergillus* endotoxin in intravenously infected rabbits in 1939, this type of experimental infection in rabbits has constituted the main body of animal models of systemic aspergillosis for more than 20 years. Later, however, rabbit models of primary pulmonary aspergillosis that more closely mimic the human disease were developed and have been used by increasing numbers of investigators for studies of pathogenetic and immunological aspects of aspergillosis.[31,77,82] The rabbit also provides the experimental model of some specific forms of aspergillosis, such as endophthalmitis and endocarditis.

2. Systemic Aspergillosis

A fatal systemic aspergillosis has been produced in rabbits administered saline suspensions of spores of *A. fumigatus* or *A. terreus* by the intravenous route.[31,83,84] Aspergillosis established in rabbits takes the acute or chronic course of infection, depending on the size of inoculum.[31] Rippon and Anderson[31] have demonstrated that a strain of *A. terreus* isolated from a patient with meningitis was as virulent as a clinical isolate of *A. fumigatus* from human pulmonary infection, but of greater virulence than that of the soil *A. terreus* isolates. There is evidence showing that the resistance of rabbits to the challenge of *Aspergillus* is profoundly lowered by immunosuppressive agents such as corticosteroids and cyclophosphamide; the treated animals developed extremely disseminated aspergillosis, in which the lungs were predominantly involved with accompanying involvement of the kidneys, liver, spleen, and central nervous system, in descending order.[31,84] Histologic findings in sections of these organs revealed multiple areas of extensive mycelial invasion with or without abscess formation in all the infected animals.[84] Rippon and Anderson[31] again reported that, irrespective of the route of inoculation, the lungs were almost always involved if the rabbits were infected at all, thus suggesting the selectively high susceptibility of the rabbit's lung to aspergillosis.

3. Primary Pulmonary Aspergillosis

The rabbit model of pulmonary aspergillosis has been generally produced by the intratracheal infection with *A. fumigatus* and, less frequently, *A. flavus* or *A. niger*.[31,77-79] Rabbits were lightly anesthetized and 10^6 to 10^8 spores suspended in 0.5 to 2 mℓ saline were injected directly into the trachea exposed through a short skin incision in the midline of the neck, basically, according to the method described by Damodaran and Chakravarty.[85] Fatal infections could be established only in the rabbits which had been treated with steroids or other immunosuppressive drugs.[31] The microbiological and histopathological studies demonstrated the primary and predominant involvement of the lung with secondary dissemination of the fungus to the liver, spleen, and kidneys and, moreover, a significant defensive role of pulmonary alveolar macrophages against injected spores or subsequently invading mycelia.[77-79] The cellular response to spores or mycelia of *Aspergillus* in the pulmonary tissue of previously sensitized rabbits were markedly enhanced, probably through activation of alveolar macrophages.[79,82]

4. Endophthalmitis

Considering the usefulness of the rabbit as a model to study hematogenous endophthalmitis caused by *Candida albicans*,[86] Fujita et al.[87] attempted to develop such a type of ocular disease in rabbits given intravenous injection of *A. fumigatus* spores. However, these workers failed to establish culture-positive, ophthalmoscopically visible endophthalmitis in any rabbits challenged with various sizes of inoculum. On the other hand, Ellison[88] and Segal et al.[89] have successfully produced experimental endophthalmitis in rabbits following a direct intravenous injection of *A. fumigatus* spores; 1 week after the inoculation, the eyes of the animals challenged with 500 fungal spores developed severe inflammatory lesions in the anterior chambers, iridis, and vitreous cavity, and a culture of the vitreous humor showed heavy growth of *A. fumigatus*. In this model, topically administered pimaricin as well as topically, systemically, or concurrently administered miconazole showed significant protective or therapeutic efficacy.[88,89]

5. Endocarditis

Carrizosa et al.[90] produced *Aspergillus* endocarditis in rabbits which had received intracardiac catheter and, 24 hr after, the inoculum of *A. fumigatus* spores by the intravenous route according to the procedures described previously.[91] The successfully infected animals developed culture-positive large occlusive vegetations on the aortic valves, with accompanying disseminated infections involving the kidneys, lungs, liver, spleen, and brain.[90] The number of viable counts of the vegetation was significantly lowered when the animals were systemically treated with amphotericin B and flucytosine, alone or in combination.[90]

D. Other Animal Models
1. Introduction

Varying species of mammals other than the mouse and rabbit, as well as avians, have been employed to induce experimental *Aspergillus* infection. Among them are several species of natural hosts, such as cows, sheep, and chicks, which have been established highly susceptible to aspergillosis. Placentitis in pregnant cattle or sheep, pulmonary infection in young chicks, and allergic bronchopulmonary aspergillosis (ABA) in rhesus monkeys are typical examples of experimental models that closely approximate the comparable human disease.

2. Rat Infection Models

Turner and his coworkers[33,92] have experimentally induced pulmonary aspergillosis and cranial aspergillosis in rats administered the inoculum of *A. fumigatus* spores by the intratracheal and intravenous routes, respectively. For the intratracheal inoculation, the spore suspension was introduced into the trachea with the aid of a catheter inserted down the throat to the junction. Significant histological involvement of the lungs was found only in the animals receiving repeated subcutaneous injection of steroids and multiple intratracheal injection of spores and, moreover, notwithstanding the severity of the lesions, the animals did not develop fatal hyphal bronchopneumonitis. It would seem, therefore, that rats may be more resistant to pulmonary infection with *Aspergillus* than mice or rabbits. Rapid and fatal cranial aspergillosis was induced in normal rats after the intravenous injection of *A. fumigatus* spores.[92] In infected animals, although the inoculated spores concentrated predominantly in the liver and lungs, and lesions developed in these tissues, only lesions in the cerebral tissue displayed the presence of hyphae.

Turner et al.[28] also infected rats via the subcutaneous and intraperitoneal routes with *A. fumigatus* spores. When infected subcutaneously, significant histological changes, particularly the formation of granulomas associated with presence of both spores and hyphae (frequently observed in mesenterial and paratracheal lymph nodes), were induced only in cortisone-treated animals. When infections were induced intraperitoneally, lesions containing hyphae were occasionally produced in the liver, spleen, and mesenteric lymph nodes, although such lesions were more frequent and extensive in cortisone-treated animals.[28]

3. Primate Models of Allergic Aspergillosis

Slavin et al.[93] have created an experimental model of ABA in rhesus monkeys. The monkeys were immunized with aerosolized preparations consisting of aqueous extract of *A. fumigatus* and killed *Aspergillus* powder by inhalation through an endotracheal tube. The monkeys, immunized with IgG precipitating antibody to *A. fumigatus*, who received human allergic serum containing IgE antibody by infusion showed an inflammatory response in the animal lung which approximated the histopathology observed in the human disease. These workers have further demonstrated that similar histopathological changes were also produced in the skin of rhesus monkeys with IgG antibody to *A. fumigatus* when the animals were injected at multiple sites intradermally with human serum rich in IgE against *A. fumigatus*.[94]

4. Cattle and Sheep Models of Placentitis

Aspergillus placentitis has been experimentally produced by intravenous inoculation of the two species of natural hosts, pregnant cow[95-98] and pregnant sheep,[99-102] with *A. fumigatus* spores. The ovine infection has been more preferably used as an experimental model of the disease for examination of the pathological and immunological changes which occur during the course of placental infection because this model appears to have a number of advantages over the use of cattle for study of mycotic placentitis.[99,101-103]

5. Avian Aspergillosis Models

Experimental avian aspergillosis has been induced in young poultry, such as the chick[104-106] and duck,[107] by intravenous injection of saline suspensions of spores or inhalation of aerosolized spores from *A. fumigatus* or *A. flavus*. After inhaling spores, the birds developed culture-positive plaques which persisted for several weeks.[104] It was also demonstrated that ducks developed considerable resistance to infection by inhalation within the first few days of life.[107,108]

II. CANDIDIASIS

A. Introduction

Candidiasis (candidosis), caused by several species of the genus *Candida*, is the most widespread and prevalent mycotic disease of man. It can be a superficial or systemic infection. Superficial candidiasis is a common infection of the skin and mucous membrane. The usual sites are major and minor skin folds, the anal, vulvovaginal and perioral mucocutaneous junctions, and the nail-bed and nails. Candidiasis in relatively healthy individuals most commonly involves superficial infections of the skin or the mucous membranes of the mouth and vagina.[109,110] Although superficial candidiasis is basically a local infection involving only a limited part of the skin and/or mucous membrane in the body, it can be the source of organisms that give rise to deep and systemic forms of infection in the compromised host. Systemic candidiasis, a serious problem predominantly of the temporarily or chronically compromised host, involves the internal organs of vital importance. The lesions can be localized in a single or in several organ systems. These include: the heart (cardiac candidiasis, in particular *Candida* endocarditis), respiratory system (bronchial and pulmonary candidiasis), central nervous system (*Candida* meningitis and cerebral candidiasis), kidney (renal candidiasis), and bone (*Candida* osteomyelitis and arthritis). However, in the compromised host, *Candida* organisms can be widely disseminated in the body through hematogenous and/or lymphatic spread resulting in more fulminating and serious forms of the disease.

Eight species of the genus *Candida* have been found to be pathogenic for man: *C. albicans, C. stellatoidea, C. tropicalis, C. parapsilosis, C. krusei, C. pseudotropicalis, C. guilliermondii,* and *C. glabrata* (formerly *Torulopsis glabrata*).[110] The most frequent causative agent of superficial and systemic candidiasis in man is *C. albicans,* which is the endogenous species in man. In man and many other animals, this yeast is present as a member of the body's natural flora on mucous membranes of the various parts of the alimentary tract from oral cavity to the intestine and vagina in 10 to 50% of healthy subjects.[111] In such habitat, *C. albicans* ordinarily lives in balance with the other members of microbial flora and merely exists there as a colonist. But various predisposing factors can upset this balance and lead not only to a local overgrowth of the organism (oral, esophageal, or enteric candidiasis), but also the most overwhelming form of candidiasis which involves a variety of internal organs through presumed extraintestinal dissemination.[112,113]

In recent years, the incidence of infections with *Candida*, in particular disseminated candidiasis, has steadily increased mainly due to the wider use of immunosuppressive drugs and broad-spectrum antibiotics.[114-117] Burn patients,[118] patients with malignancies,[113,119,120] those in immunodeficient states,[121,122] or those undergoing therapeutic manipulations, such as surgery[123] and intravenous hyperalimentation,[124] are also predisposed to candidiasis. The variety of predisposing factors and clinical forms of the disease as well as the poor diagnostic method and antifungal chemotherapy now available have made candidiasis a serious clinical problem. This situation has led us to the need for better understanding of the immunity, host-pathogen interactions, and pathogenesis involved in human candidiasis, and that of developing more effective and reliable methods for controls of the disease. The valid and reproducible animal model that mimics the comparable form of the human disease should be prerequisite for such studies.

In the past two decades, increasing numbers of animal models of candidiasis produced in a variety of laboratory animals, as well as in some domestic animals, have been described in the literature. In experimental models of candidiasis, with very few exceptions, *C. albicans* has been the only species used as the challenge organism for producing candidiasis, because of its medical importance and virulence for almost all laboratory animals. Researchers in the pharmaceutical industries have made a substantial contribution to developing those types of animal models which can be useful for preclinical evaluation of antifungal regimens.[125]

B. Animal Models of Systemic Candidiasis
1. Introduction

Fatal, naturally-occurring systemic candidiasis has been reported in capuchin monkeys,[126] but it has not been recognized in any species of laboratory mammals in use. Experimental models of systemic candidiasis have been produced in the rabbit, rat, guinea pig, and, most often, in the mouse. In particular, the mouse has been providing the best available model for studying various aspects of *Candida* and candidiasis. In these models, the most favorably used route of inoculation has been the intravenous route, which can solidly establish the fatal disease without the aid of immunosuppressive treatment of the animals, in challenge doses significantly lower than those for the intraperitoneal route or others. Irrespective of the route of inoculation and the species of animals used, once systemic candidiasis is established in the inoculated animals, the course of the infection is surprisingly similar among a number of different models thus far developed. All the cultural and histopathological studies have demonstrated that the kidney is, unexceptionally, the major target of infection with accompanying less severe involvement of the liver, spleen, and heart.[127] In addition, several forms of local manifestation, such as endogenous endophthalmitis and cutaneous candidiasis occurring through hematogenous dissemination of *Candida* organisms from the primarily affected organs have been recognized in certain models.

2. Intravenous and Intraperitoneal Infection
a. Murine Models
i. Variance in Virulence of *C. albicans* Strains

In almost all cases, *C. albicans* strains that were originally isolated from patients with candidiasis have been used to prepare an inoculum for a challenge of animals. The determinants of virulence for mice among strains of *C. albicans* have not been completely defined, although a number of factors, including inducible proteases, phospholipases, *Candida* endotoxin, and the capacity to form hyphae, have been proposed.[110] Saltarelli and investigators[128] have made a comparative study of the virulence of *C. albicans* morphological mutant strains to relate chlamydospore production and germtube formation in mice receiving an intraperitoneal challenge of the inoculum and observed that the mycelial strains were not more lethal than the yeast-like strains, and that neither chlamydospore production nor germtube formation was related to the virulence for mice. On the other hand, Evans and Mardon[129] have provided evidence suggesting that disseminating systemic *C. albicans* infections were more likely to be initiated by the yeast-like form of the organism and that that may play a vital role in innate host resistance to hematogenous dissemination of *C. albicans.* Besides, Saltarelli et al.[128] could not find any relationship between the proteolytic activity and virulence in their strains. This result contradicts the study of Staib[130] who observed that *C. albicans* strains capable of lyzing albumin in agar were highly virulent for mice. Notwithstanding the possible variety of biological traits of *C. albicans* strains so far used in murine models, the literature tells us that all the authors have successfully produced fatal infections in mice by injecting 10^5 to 10^7 viable yeast cells of *C. albicans* intravenously. Plempel[131] has tested the virulence of 36 *C. albicans* strains freshly isolated from clinical materials for CF1-SPF mice following an intravenous inoculation of 6 to 10×10^5 organisms. He found that 35 of 36 strains were similarly pathogenic for mice; 60 to 100% of mice infected with either of the virulent strains succumbed in 6 days after challenge.

ii. Susceptibility to *C. albicans* Infection of Different Murine Strains

The use of various inbred strains of mice to explore the immunological and genetic factors involved in determining resistance to various infectious agents has been used successfully in recent years. Bistoni et al.[132] have studied the susceptibility of different mouse strains to infection with graded numbers of *C. albicans* injected intravenously. Inbred C3H, BALB/c, and hybrid CD2F1 mice showed approximately similar degrees of susceptibility to the infection, whereas C57Bl/6 mice were slightly more resistant. The relatively high resistance of mice from other C57Bl substrains to an intravenous challenge with *C. albicans* have also been confirmed by Neta and Salivan.[133] They found that alloxan-induced diabetic mice of both C57Bl/10SNJ and C57Bl/NsJ were infection resistant in that *C. albicans* administered intravenously in a dose of 4×10^4 cells was not cultured from kidney 14 to 21 days after challenge.

The congenitally athymic (nude) mice and New Zealand Black (NZB) mice have been favored to study the role of thymus-dependent, cell-mediated immunity in disseminated candidiasis. Rogers et al.[134] have demonstrated that nude mice have a greater capacity than their phenotypically normal littermates to prevent growth of intravenously administered *C. albicans* in the kidneys and to clear the organisms from the liver. Similarly, mice of the NZB strain, which spontaneously develop a selective deficiency of cell-mediated immune functions, showed a greater susceptibility to lethal infection with *C. albicans*, as well as with *Aspergillus fumigatus* and *Cryptococcus neoformans*, when challenged intravenously.[135]

The important role of a humoral-factor complement that plays in host defense against experimental *C. albicans* infection was demonstrated in the work of Merelli and Rosenberg.[136] They found that the susceptibility to the infection of complement-positive CF1 mice was comparable to that of complement-deficient mice of the same strain. This line of studies was extended more recently by Hector et al.[137] who introduced several inbred strains of mice. These investigators examined mice from six genetically distinct strains for their susceptibility to *C. albicans* when challenged with several different doses of inoculum. Mice from the six groups showed substantial differences in resistance to challenge based on mortalities and quantitative cultures of kidneys; mice from strains C57Bl/6J and BALB/cByJ showed the least susceptibility; mice from strains C57Bl/HeJ and CBA/J, moderate susceptibility, and mice from strain DBA/2J, the highest degree of susceptibility to challenge. These results suggest a major role for complement in the innate resistance to *C. albicans* in the murine model because DBA/2J and A/J strains have been shown to be deficient in the C5 component of complement.[138] Thus the studies of Hector and co-workers confirmed, not only those of Morelli and Rosenberg,[136] but also Gelfand et al.[139] who used guinea pigs, in which animals lacking an intact alternate pathway of complement did not survive as long as animals with an intact complement system.

Sex difference in resistance to *C. albicans* infection, as well as in immune response to this infection, both favoring the female, have been reported in various experimental animals including mice.[128,140,141] However, contradictory results were obtained by Rogers and Balish[142] who studied systemic candidiasis in mice and germ-free rats given fatal doses of *C. albicans* by both intraperitoneal and intravenous routes.

iii. Predisposing Factors

Experimental models of systemic candidiasis, in particular those established in mice by intravenous challenge with *C. albicans*, have been utilized to study the effects of a variety of immunosuppressive regimens on the susceptibility of the host animal to infection with the microorganism. Increased susceptibility to the infection has been reported in X-irradiated mice,[143,144] corticosteroid-treated mice,[143,145] cyclophosphamide-induced granulocytopenic mice,[146-149] and in mice treated with other drugs such as azathioprine.[150] Cyclophosphamide has been also used to study the protective effect of immunoactive substances on *C. albicans* infection in severely immunosuppressed mice[132,151,152] and to assess the role of several immunological responses in the acquired resistance.[153] One of the experimental models to study the basis for enhanced susceptibility of diabetic animals to infections, such as those by *C. albicans*, includes infection of mice following administration of alloxan. Salivan and Turner[154] found that mice became more susceptible to *Candida* infection when they were given a diabetic dose of alloxan 10 days before inoculation of the yeast.

Host resistance to *C. albicans* in a model tumor system has been studied by Robinette and Mardon.[155,156] They demonstrated that the lethal response to *C. albicans* infection was significantly delayed in (C57Bl × BBA/2)F1 mice inoculated intravenously with the organism 6 to 16 days after transplantation of Lewis lung carcinoma as compared with noncancerous control mice, probably because of the enhanced microbial destruction in fixed reticuloendothelial-system cells of the tumor-bearing mice.

iv. Applications

Since fatal infection of mice with *C. albicans* is readily producible in a large scale and can provide a reproducible and quantifiable model of systemic candidiasis, it has been most frequently used among various animal models of the disease to evaluate the protective role or the effect of immune responses in the animal, several biological and chemical immunomodulators,[157-161] or the therapeutic effect of currently available antifungal drugs.[131,148,150,162-172] In these experimental systems produced in mice with or without immunosuppressive treatment, significant protective or therapeutic efficacy of systemically administered, currently available, or promising antifungal agents, such as amphotericin B, flucytocytine, miconazole, ketoconazole, BAY n 7133, and itraconazole, which were used alone or in appropriate combinations or by means of a new drug delivery system (e.g., liposomes), have been reported.

b. Rabbit Models

Next to the mouse, the rabbit has been most frequently used as a model of systemic *C. albicans* infection. Rabbit models of candidiasis are especially useful for such studies as those that follow the titer of *Candida* antigens or antibodies to *Candida* in the serum of animals in the course of infection because of their larger size which enables us to obtain adequate amounts of blood samples repeatedly over the experimental period. On a weight basis, rabbits appear to be more susceptible to *C. albicans* infection than mice. Intravenous challenge of rabbits with an inoculum of the organism in doses of 5×10^5 to 10^7 have been reported to induce rapid fatal disseminated candidiasis; all animals died within 10 days after infection.[173-177]

The kidney was always the major target of infection and the liver and spleen were less severely affected. Rabbits given lesser doses of challenge inoculum developed non-fatal chronic systemic infection where *C. albicans* were recovered from the kidney, liver, and spleen over a 10 week or longer observation period.[173] Very recently, Repentigny et al.[178] produced disseminated candidiasis in cortisone-treated rabbits following intravenous injection of 10^7 *C. albicans* organisms. All animals had either cultural and/or histological evidence of invasion by *C. albicans* in their kidneys and livers. The survival of the animals was prolonged compared with that observed in most similar studies on detection of serum mannans,[179-182] arabinitol,[176,183] and mannose[184] in experimental candidiasis. Thus Repentigny et al.[178] claimed that this experimental model in rabbits receiving cortisone may more closely resemble the often occurring subacute type of candidiasis in humans.

c. Rat Models

Systemic infection of rats has been induced after a challenge of the animals with 10^6 to 10^7 *C. albicans* organisms by the intravenous route. These challenge doses were fatal to 95% or more of the animals within 2 weeks post challenge.[184-186] Balk et al.[186] described details of symptomatology and histology of rats infected with fatal doses of *C. albicans*. As is the case for other experimental animals, the kidney was the most severely diseased organ, from which *C. albicans* were cultured. Virtually the same results have been obtained in germ-free rats.[142] Balk et al.[186] claimed that their model of disseminated candidiasis in laboratory rats showed clinical and histological changes similar to those reported for other animals and humans dying of this disease.[187,188] These investigators[186] and Galgiani and Van Wyck[185] used this rat model for in vivo evaluation of antifungal drugs.

d. Guinea Pig Models

Guinea pigs appear to be considerable resistant to *C. albicans* infection irrespective of the route of inoculation.[127] The earlier report of Winner[189] described fatal *C. albicans* infection produced in guinea pigs after intravenous injection of the inoculum. More recently, Hurley and Fauci[190] developed a model of disseminated candidiasis in guinea pigs injected with the lethal doses of *C. albicans* via an intravenous or intraperitoneal route. These workers stressed the similarity of their model to human candidiasis in histological and other aspects. Gelfand et al.[139] used this guinea pig model, produced in normal animals, animals congenitally deficient in component C4 of the complement (C4D animals), and animals depleted of the alternate pathway as well as component C3-C9 of the complement sequence by treatment with cobra venom factor (CVF), to study the role of complement and its various pathways of activation in the early stage of candidiasis. As a result, they found that mortality of infected animals and also infection of the kidney was markedly enhanced in the CVF-treated guinea pigs but not in C4D animals, when compared with untreated normal animals. Very recently, Van Cutsem and investigators[191-193] have developed experimental systemic candidiasis in guinea pigs given *C. albicans* inocula via an intravenous route and described microbiological, hematological, and histopathological aspects of their model. They claimed that the systemic *C. albicans* infection thus established was characterized by not only involvement of the various internal organs such as the kidney, liver, brain, eye, and stomach, but also dissemination to the skin.

e. Canine Models

Ruthe et al.[194] induced systemic candidiasis in cyclophosphamide-induced neutropenic canines after intravenous challenge with *C. albicans*. A normal dog could tolerate up to 10^7 *C. albicans* organisms administered intravenously, while 10^6 organisms invariably resulted in widespread disseminated candidiasis in leukopenic animals. Using the canine model produced in leukopenic dogs, this group of investigators has studied the therapeutic effect of granulocyte transfusion on the severity of candidiasis.[194-196]

f. Comparison Between Intravenous and Intraperitoneal Routes

The intraperitoneal route of inoculation has been used, although much less frequently than the intravenous route, to produce experimental infection with *C. albicans* in mice,[126] rabbits,[197,198] and guinea pigs.[127,189,190] In all these species of laboratory animals, this route of inoculation appears to be more efficient in establishing fatal systemic candidiasis than the intravenous route. Hurley and Fauci[190] have infected guinea pigs using both intravenous and intraperitoneal routes of infection; all the animals challenged intravenously with the maximal dose of inoculum (10^8 CFU) of *C. albicans* succumbed to the infection within 24 hr. The use of the graded doses of inoculum determined the 50% lethal dose as 3×10^6 CFU. On the other hand, no mortality or apparent illness resulted in the animals receiving this maximal dose of inoculum, although 2-weeks postchallenge autopsies of these apparently healthy animals revealed multiple sites of abscess formation in the omentum and the serosal surfaces of the abdomen and liver, but no viable organisms could be cultured. As compared with other rodents of laboratory use, guinea pigs appear to be particularly resistant to intraperitoneal challenge with *C. albicans*. This may be associated with the high fungicidal efficacy of peritoneal macrophages of guinea pigs.[199]

Bayer et al.[197] developed a rabbit model of peritoneal candidiasis (candidial peritonitis) in the animal with a challenge of an inocula containing 10^{10} *C. albicans* cells. These investigators noted the relative refractoriness of the rabbit peritoneum to such large inocula of *C. albicans* as was seen in guinea pigs.[190] Intraperitoneal injection of 10^8 to 10^9 organisms were needed to induce intra-abdominal infection in rabbits. In contrast, 10^5 CFU of the same *Candida* strain, given by intravenous injection, caused disseminated candidiasis in nearly all rabbits.[200] Even though intraperitoneal *C. albicans* could be established, there was a tendency for gradual resolution of peritoneal lesions over 2 to 3 weeks postinfection. However, approximately half of the infected animals developed an ocular manifestation, endophthalmitis, through hematogenous dissemination.[198]

3. Intracardiac and Intraarterial Infection

Very infrequently, an intracardiac route[201] and an intra-arterial route[202,203] of inoculation has been used to induce systemic *C. albicans* infection in rabbits and rats, respectively. The latter rat model may be useful for studying cerebral involvement in deep candidiasis. Parker et al.[202,203] established deep candidiasis in healthy uncompromised Wistar rats by injecting *C. albicans* into the right internal carotid artery. The infected animals succumbed to the disease within 7 days after a challenge with 10^7 organisms. The kidney was most severely infected, and the entire brain was diseased without involvement of leptomeninges. Parker and workers claimed that their rodent model is similar to human cerebral candidiasis.[202]

4. Local Manifestations

a. Ocular Manifestation (Endogenous Candida Endophthalmitis)

The first model of endogenous *Candida* endophthalmitis was produced in rabbits by Hoffman and Waubke in 1961 by intravenous injection of *C. albicans*.[204,205] Yeast cells were detected in choroidal vessels 15 minutes following injection and progressed to the outer retina within 1 hr. Pseudomycelia developed as the organism invaded the retina.

The disease progressed to typical stages of vitreoretinal nodules, multifocal vitreous abscesses, and destructive endophthalmitis. The ocular lesions may persist for months and some lesions may heal spontaneously. Vergara et al.[206] have also described the model of endogenous endophthalmitis produced in rabbits given intravenous injections of *C. albicans*. Santos and investigators[207] produced endophthalmitis in pigmented rabbits by intravenous injection of 3×10^6 *C. albicans* organisms to test the efficacy of laser photocoagulation. Retinal lesions developed and progressed in all animals infected within 4 days after inoculation. Kidney and brain cultures were uniformly positive. In more recent years, Edwards and investigators[200,201] developed a new rabbit model of hematogenous endophthalmitis with an intravenous challenge of 10^5 *C. albicans* organisms. In this model, retinal lesions were detected ophthalmoscopically 3 days following challenge and persisted for 80 days; 88% of the rabbits had lesions with clinical signs identical to that of human lesions. A similar rabbit model has been reported by Demant and Easterbrook.[208] Using the Edwards and co-workers' model, studies have been performed on the ultrastructural figures of *Candida* endophthalmitis in rabbits[209] and on evaluation of anterior chamber aspiration as a mycological diagnostic method.[210] Investigators from the same laboratory also studied the ocular pathogenicity of several different species of the genus *Candida*, including *C. albicans*, *C. guilliermondii*, *C. krusei*, *C. parapsilosis*, *C. stellatoidea*, and *C. glabrata*. The results showed a relative resistance of ocular tissues to hematogenous *Candida* infection with species other than *C. albicans*.[211,212]

Therapeutic trials of experimental *C. albicans* endophthalmitis have been made by Rimbaud and others[213] and Jones and others.[214,215] The former group of workers produced typical retinal lesions in rabbits given 5×10^5 *C. albicans* organisms per kilogram. With this model, they proved the therapeutic effects of intravenously administered amphotericin B.[273] Jones established progressive endogenous *C. albicans* endophthalmitis challenged intravenously with 2 to 5×10^5 organisms per kilogram, with the severity of infection being directly related to the fungal strain and inoculum size. His model was workable for evaluation of various systemic antifungal agents, such as amphotericin B, ketoconazole, and miconazole.[214,215]

In addition to the intravenous route, the intraperitoneal route of inoculation with a larger inoculum (e.g. 5×10^8 CFU) has been reported to develop hematogenous *Candida* endophthalmitis in rabbits that were significantly responsive to the therapeutic doses of amphotericin B as well.[198]

b. Cutaneous Manifestation

Development of cutaneous lesions in guinea pigs infected intravenously with fatal doses of *C. albicans* has been recognized and described by Van Cutsem and investigators.[125,191-193] Viable *C. albicans* was recovered from skin tissues with lesions 35 days postinfection.[193] Infected animals developed small red papular eruptions on the skin over the entire body. When animals were infected with sublethal doses of *C. albicans*, such skin eruptions were observed more prominently. Skin lesions might result from passage of the organisms disseminating through the basement membrane of the epidermis to the superficial horny layers of the skin.[125]

c. Other Superficial Manifestations

Thienpont et al.[125] have reported that female guinea pigs infected intravenously with *C. albicans* developed *Candida* vaginitis. When pregnant rodents (rats, guinea pigs, and mice) were administered *C. albicans* via an intravenous route, the organisms disseminated through the placenta to the fetus in the uterus which resulted in extensive hyphal growth on the skin of unborn fetuses harbored in the infected mother animals.

In the chicken and some other avian species, an intravenous challenge of *C. albicans* resulted in systemic infection with cutaneous lesions as observed in guinea pigs.[125] In infected cocks, lesions were produced on combs and wattles. These investigators also described gastrointestinal involvement following systemic infections experimentally induced in various species of animals.

C. Animal Models of Localized Candidiasis
1. Introduction
Notwithstanding the size of challenge inoculum, the route of inoculation, the species, or the immunological status of the animals used, *C. albicans* introduced systemically into animals via either intravenous route, intraperitoneal route, or even intraarterial route usually produced infections that were mostly confined to the kidney with slight degrees of involvement of the liver, spleen, and, less frequently, eye as mentioned in the preceding section. In the literature, experimental models of varying forms of candidiasis other than disseminated forms have been described. Appropriate procedures for inoculation or surgical pretreatment of the animal should make it possible to induce such specific types of *Candida* infection, in particular those localized in certain tissues or organs of the animal.

2. Exogenous Endophthalmitis
Exogenous *Candida* endophthalmitis has been developed mainly for the purpose of providing an animal model which is valid for evaluation of antifungal agents administered intravitreously. Axelrod and Peyman[216] have produced progressive vitreous abscesses in rabbits by intravitreous injection of 10^3 *C. albicans* organisms. Vitreous clouding was evident in all animals within 24 hr post-inoculation. With this model, protective or therapeutic efficacy of intravitreous amphotericin B and its methyl ester, used alone or in combination with vitrectomy, has been demonstrated.[216-219]

3. Cutaneous Infections
Experimental acute cutaneous infections with *C. albicans* have been produced in man[220-222] as well as in a variety of animals: birds such as chickens, turkey and guinea fowls[224,225] dogs;[226] rabbits;[227,228] guinea pigs;[125,229,230] rats, and mice.[231-233] In some cases, animals were treated with subcutaneous or topical cortisone prior to inoculation.[227] Dogs,[226] guinea pigs,[227] rabbits,[228] rats, and mice[231] have offered successful models of acute cutaneous disease when *C. albicans* or related *Candida* species were inoculated epidermally under occlusive dressings. Ray and investigators[231] developed the occlusive model produced in newborn rats and mice. In the infected rodents, development of microscopic pastules, similar to the lesions which occur in men, was observed only at sites of *Candida* hyphal invasion.[231,234] These investigators also demonstrated that of six *Candida* species tested, only *C. albicans* and *C. stellatoidea* could induce subcorneal microabceses when inoculum of each *Candida* species was applied epicutaneously and the area was occluded with a dressing.[234,235]

Giger et al.[233] have produced cutaneous candidiasis model in mice without occlusive dressings. These workers injected suitable doses (5 to 10 × 10^5 cells) of *C. albicans* intradermally into the shaved skin of the animals, which then developed a lesion that was confined to the dermis and was of a limited duration (2 to 3 weeks). The lesions were essentially abscesses containing large numbers of PMN. Domer and Moser[236] have used this model of self-limiting dermal infection for studies of protective immune responses to *C. albicans*. A similar cutaneous model was also produced in rabbits following intradermal injection of *C. albicans* inoculum.[201,237]

Among species of rodents, the guinea pig appears to be relatively susceptible to cutaneous *C. albicans* infection.[125] Sohnle and co-workers[230,238,239] produced cutaneous candidiasis in guinea pigs either by using occlusive dressings over the microorganisms or by applying them to shaved skin directly without occlusive dressings. In either model, there was clearance of the infecting organisms from the skin by a process involving profuse scaling of the keratinized layer in which they were confined. The latter type of guinea pig model was also produced by Van Cutsem and Thienpont.[229] They also found that treatment of the animals with alloxan was effective in establishing persistent infection. Using this cutaneous model produced in alloxan-diabetic guinea pigs, Cauwenbergh et al.[240] confirmed the therapeutic efficacy of topically administered ketoconazole.

4. Subcutaneous Infection

Huang et al.[177] infected rabbits by subcutaneous injection, which was repeated 7 times during a period of 16 days, of *C. albicans* organisms. The rabbits showed evidence of local infection as determined by induration and redness at the injection sites. Grunberg et al.[170] produced experimental *C. albicans* infection in cortisone-treated mice given subcutaneous injection of the organism in order to evaluate the antifungal effect of flucytosine in vivo.

5. Muscular Infection ("Mouse-Thigh Lesion")

Selbie and O'Grady[241] and O'Grady[242] originally described a tuberculous thigh lesion produced in mice by intramuscular injection of *C. albicans* organisms. Thompson[243] used this model to study the effects of a number of chemotherapeutics and corticosteroids on thigh lesions. Since such a model of self-limiting infection with *C. albicans* looked to be highly reproducible and quantifiable, Pearsall and Lagunoff[244] later detailed the method to establish the mouse-thigh lesion model which may permit a long-term study of immunological responses to *C. albicans* infection. These investigators injected a standard inoculum of 5×10^8 organisms into the thigh muscle of C57Bl/Ks mice and produced an easily measurable thigh lesion that was self-limiting by 4 to 6 weeks. This model has been favorably used by these and other investigators for immunological studies.[245-247]

During the course of the study, Pearsall and Lagunoff[244] unexpectedly found that *C. albicans* infection induced amyloidosis in infected mice and that amyloidosis was induced not only by viable cells of *C. albicans*, but also by both viable *Saccharomyces cerevisiae* cells and heat-killed *C. albicans* in mice of susceptible strains (e.g., C57Bl/Ks). Mann and Blank[248] described a method of inducing experimental amyloidosis in mice using intramuscular injection of lyophilized *C. albicans* cells. Savage and Tribe[249] compared the two methods of inducing amyloidosis and found that once amyloidosis was induced, the disease always progressed until the animals died of generalized amyloidosis. Thus, this experimental model of amyloidosis should be suitable to use in studying pathogenesis of amyloidosis and assessing the treatment of this disease.

6. Arthritis

With inoculation of 10^8 *C. albicans* yeast cells into the supra-patellar bursae of rabbits, septic synovitis was induced.[250] In the infected rabbits, effusion appeared in the second week after inoculation and maximally developed and became purulent in the third and fourth week, thereafter subsiding to a chronic pannus with cartilage destruction and repair within 8 weeks.

7. Vaginitis

Experimental model of *Candida* vaginitis should have the validity for studying the effect of antifungal regimens for the disease in humans. In this view, relatively simple, rapid, and reproducible procedures for inducing *Candida* vaginitis in experimental animals have been devised by a number of investigators. Since the role of the hormonal milieu is evident by the resistance of oophorectomized, nonestrogen-supplemented animals to *C. albicans* infection,[251] oophorectomy and hormonal maintenance is strictly required to maintain *Candida* infection in the vagina of experimental animals. So far as known, almost all workable models of *Candida* vaginitis have been produced in oophorectomized rats that were kept in estrus before and after one or two intravaginal challenges with *C. albicans* by weekly injection of estrogens.[252-254] Establishment of acute vaginal candidiasis in inoculated rats can be verified by cultural studies, as well as histological studies which should show extensive desquamation of the vaginal epithelia.[125,252,255,256] *Candida* vaginitis models produced in oophorectomized rats have often been used to evaluate various antifungal agents used topically (e.g., flucytosine, polyene antibiotics, econazole, miconazole) or orally (e.g. ketoconazole, BAY n 7133, itraconazole).[125,172,251,252,254-258] Very recently, more chronic *C. albicans* vaginitis in oophorectomized rats has been developed by Sobel and Muller.[259]

Very few attempts have been made to produce *Candida* vaginitis in animals other than rats. Segal et al.[260] have induced vaginal infection with *C. albicans* in mice by inoculating the animals with the organisms during the estrus stage of the mouse hormonal cycle. The infection was self-resolving with very few fungal elements being observed microscopically on the 3rd to 4th day postinoculation.

8. Cystitis

Eng et al.[261] reported having produced cystitis in rabbits by using a tube filled with *C. albicans* inoculum. However, there is no description of the course or histopathology of the infection produced.

9. Oral Infection

In 1936, Mackinnon[262] first reported that he could induce oral candidiasis in one rabbit by oral administration of *C. albicans* organisms. Later studies showed that Wistar rats may be a more suitable model of oral candidiasis. More recently both short- and long-term oral candidiasis has been established in adult rats given inoculation of *C. albicans*. Inoculation can be performed by placing *C. albicans* cell suspensions in the mouth of rats. Intensive studies on the rat model of oral candidiasis have revealed that the disease was not more readily induced in weanling rats than in adult rats, and that both a carbohydrate-rich diet and tetracycline medication favored the carriage and, perhaps, the infectivity of *C. albicans* in the oral cavity of conventional or conventionalized SPF-rats as well as germ-free rats.[263-270] In this rat model, definite *C. albicans* strain-related differences in pathogenicity for the lingual mucosa of the animals was demonstrated.[271]

Birds also should provide a useful model of oral candidiasis because of their high susceptibility to *C. albicans* infection. Naturally occurring turkey crop candidiasis has been well-recognized. Balish and Phillips[272] has succeeded in infecting the crops of gnotobiotic chicks with the oral-route inoculation of *C. albicans*. Thienpont et al.[125] have induced *C. albicans* infection of the crop and esophagus in several avian species, such as the turkey, guinea fowl, and chick, following the oral route of inoculation. The infection frequently resulted in dissemination to the mouth, tongue, interlabial space, and stomach. Phillips and Balish[273] attempted to infect mice by oral challenge with *C. albicans*, but they found that the candidiasis produced was confined to the stomach.

10. Gastrointestinal Infection

It is well-established that *C. albicans* can colonize the gastrointestinal tract in germ-free mice and chickens.[277,278] However, *C. albicans* could not compete with either the bacterial flora in the gut of normal mice nor with *Escherichia coli* inoculated together in germ-free mice.[276] Thus conventional mice orally inoculated with *C. albicans* are unable to maintain this organism in the intestinal tract without compromising treatment.[277,278] In this view, models of invasive gastrointestinal candidiasis should include broad-spectrum antibiotics and cytotoxic treatment which can alter the normal bacterial flora and damage the mucosal barrier. These alterations may enhance *Candida* colonization on the mucosa and allow microinvasion with occasional systemic dissemination. In the literature several successful murine models of gastrointestinal candidiasis produced in normal adult mice by oral administration of inocula using a gastric gavage have been reported.[277,279-283] In all these cases, animals were previously compromised with treatment with broad-spectrum β-lactam (e.g., ampicillin), aminoglycoside (e.g., neomycin, tobramycin, gentamicin), cortisone, estradiol and X-irradiation, which were used alone or in some combinations. The histological findings reported in these murine models revealed ulceration and necrosis, but the extent of inflammatory response was unclear.[281] Treatment of mice inoculated with amphotericin B, nystatin, flucytosine, or ketoconazole administered orally resulted in a reduction in the *C. albicans* counts in the feces.[125,277] This may suggest the usefulness of this murine model in evaluating the antifungal agents.

Different from the case with normal adult mice, infant mice appear to be highly susceptible to intragastric challenge with *C. albicans*. Pope et al.[284] and Field et al.[285] have successfully infected infant mice with the intragastric inoculation of *C. albicans*. The inoculation resulted in not only a long-term colonization of the gastrointestinal tract, but also systemic spread and lethality. The infant-mouse model may have the advantage of no requirement of compromising treatment to establish colonization in the gut. In addition, certain strains of *C. albicans* was found to persist in the gastrointestinal tract of the animals into ages at which they were normally resistant to colonization.[285] The infant-mouse model has also proven useful for studying host-pathogen interactions and efficacy of antifungal regimens.

Experimental models of gastrointestinal candidiasis have been reported in compromised rats.[281,286] Burke and Gracey[287] challenged rats with *C. albicans* by the repeated intragastric inoculation and observed significant growth of the organism in the stomach and duodenum as well as intestinal dysfunction in the infected animals.

11. Pulmonary Infection

Experimental pulmonary candidiasis that was produced by intratracheal inoculation of *C. albicans* has been reported in rabbits,[288-290] guinea pigs,[292-296] and, less frequently, in mice.[297] In the intratracheally infected rabbits, lesions were produced in the lungs,[290,298] but they appeared to be quickly and completely resolved,[291] suggesting that the rabbit lung is highly resistant to *C. albicans* infection. Kurotchkin and Lim[288] obtained lung lesions only after sensitizing the rabbits previously with the killed organism. Relatively high resistance of the lung to *Candida* infection has also been demonstrated in mice. Nugent and Onofrio[297] reported that mice uniformly survived intratracheally administered *C. albicans* inocula which killed the majority of mice inoculated intravenously.

12. Cerebral Infection

Intra-arterial inoculation of *C. albicans* into the internal carotid artery of adult rats could not produce diffuse leptomeningitis in the animals.[202] Instead, when the organism was injected into the cisterna magna of rats, *C. albicans* infection confined to the leptomeninges were established.[203] In this meningitis model, the organism was recovered only from the central nervous system.

13. Endocarditis

Experimental endocarditis caused by *C. albicans* has been induced in rabbits which received a catheter inserted via the carotid artery to the aortic valve.[299,300] This treatment resulted in a sterile thrombotic endocarditis with development of vegetations. When the animals were inoculated with *C. albicans* (approximately 10^8 viable cells) via an intravenous route, *Candida* infection of the sterile vegetations and dissemination to the other organs, especially to the kidney, were produced 4 days after inoculation.[299-301] This infection can be fatal and the mean survival time of the infected rabbits was reported to be 26 days.[302] The rabbit model of *Candida* endocarditis thus produced could afford a means of studying pathogensis[303] and of evaluating therapeutic regimens for the disease.[304]

III. CRYPTOCOCCOSIS

A. Introduction

Cryptococcosis is an infection caused by the soil-inhabiting encapsulated yeast *Cryptococcus neoformans*. It has recently been recognized that *C. neoformans* is a species whose sexual state reveals it to be a basidiomycete; the perfect state is either *Filobasidiella neoforms* or *F. bacillispora*.[305,306]* These *Filobasidiella* species have only rarely been isolated from the natural sources and the involvement of the sexual form in producing human infection is unknown.

C. neoformans is widespread in nature throughout the world. It is most abundant in pigeon habitats and this interrelationship has been confirmed many times since Emmons recovered several *C. neoformans* isolates in the course of a soil survey for *Histoplasma capsulatum* in the U.S. in 1951.[307] *C. neoformans* is also known to be resistant to dessication.[308] Dried pigeon excreta and soil in its vicinity remain the most important environmental sources of *C. neoformans*. The organism reproduces when climatic and nutritional conditions are satisfactory and may be disseminated in infectious aerosols which gain entrance through inhalation. This may result in asymptomatic or symptomatic cryptococcosis in humans and other animals. It is generally considered that the infectious material enters via the respiratory tract,[309-311] with the lungs being the first focal point of infection.[309,312-316]

Clinical and surgical evidence has been offered that supported the occurrence of pulmonary cryptococcosis.[309,315,317] Usually this is the mild and transitory phase of cryptococcal infection that probably precedes the systemic form of the disease.[318] For some reasons that are not understood, *C. neoformans* has a predilection for the central nervous system (CNS) and frequently disseminates to the CNS from apparent or inapparent primary foci in the lung through hematogenous or lymphatic spread and is manifested as the cerebral form of the disease or meningitis.[312,318,319] Occasionally, cutaneous, mucocutaneous, osseus, ocular, and visceral forms of the disease develop through dissemination from the primary pulmonary focus.

Like other infections caused by opportunistic fungi, cryptococcal infection is usually associated with some predisposing condition which has modified the immunological competence of the host. This may be a naturally developing primary or secondary

*   Now, they are replaced by *Filobasidiella neoformans* var. *neoformans* and *F. neoformans* var. *bacillispora*, respectively.

immunodeficient state or the result of administration of therapeutic agents. In man the most commonly recognized predisposed factors are leukemia/carcinoma, lymphoma (mainly Hodgkin's disease), collagen vascular disease (such as systemic lupus erythematosus), sarcoidosis, diabetes mellitus, congenital immunodeficient states; therapy with antibiotics, corticosteroids, or immunosuppressive agents (especially in malignant conditions), and renal transplantation.[318,320-322]

In addition to human infections, cryptococcosis is well-recognized as occurring naturally in diverse species such as cats, dogs, pigs, horses, cows, monkeys and cheetahs, causing CNS, pulmonary, or disseminated disease.[323] More recently, cryptococcosis has also been seen in certain wild animals from the ferret to the koala bear. In animal species other than the man, a predisposing condition is often not identifiable, although the sporadic nature of the infections suggests that individual factors play an important role in pathogenesis. Studies of experimental cryptococcosis in animals was started at the end of the 19th century, immediately after the first human cases of cryptococcosis was described by Busse. Unfortunately, most of the natural host animals are inconvenient for use in the laboratory and, instead, several species of laboratory animals including the mouse, rabbit, guinea pig, and rat have been used as an experimental model for cryptococcosis. Mice are highly susceptible to *C. neoformans*, enough to develop easily the fatal disease and, mainly for this reason, they have been most frequently used to produce experimental cryptococcal infections. Mice have offered a preferable model of systemic cryptococcosis for studying several important aspects of cryptococcosis, in particular those pertaining to pathogenesis, immunity, and chemotherapy.[324] Some other species of laboratory animals have been useful for creating rather localized forms of the disease.

B. Murine Models of Systemic Cryptococcosis
1. Introduction

There are few animal models for the study of systemic cryptococcosis. Among them, the murine model has been most frequently and preferably used for several reasons: (1) mice are highly susceptible to *C. neoformans*, (2) murine cryptococcosis is similar to the disease in humans in many aspects,[309,318,319,325-335] (3) mice are relatively easy to handle and can be used in greater numbers than larger animals, and (4) a great bank of knowledge of immunological features and immunogenetics has been established for various mouse strains. Fatal cryptococcal infection has been produced by using various routes of inoculation, such as intracerebral (intracranial), intravenous, and intraperitoneal routes. Kong and Levine[336] reported that 50% lethal doses of *C. neoformans* in inoculated mice increased in this order. Each of these three routes of inoculation is sufficiently effective for inducing systemic cryptococcal infection and the resultant fatal cryptococcosis, in particular following intravenous challenge, has been used by large numbers of investigators as a valid model for studies of virulence of various *C. neoformans* strains and pathogenesis and immunology of cryptococcosis, as well as for evaluation of newer anticryptococcal treatment regimens. However, intracerebral, intravenous, and intraperitoneal routes are all artificial routes of infection and the resulting disease may or may not simulate cryptococcosis acquired by a natural route of infection. Pappagianis et al.[337] demonstrated that each of the three artificial routes of infection allowed rapid dissemination of *C. neoformans* to the internal organs. However, when the lungs were the portal of entry, primary pulmonary disease was initiated and dissemination to the internal organs was delayed. Smith et al.[335] have observed that mice infected by the inhalation of aerosols containing high concentrations of cryptococci developed progressive cryptococcosis which was analogous to the human disease. This line of evidence offers the rationale that this route of inoculation is more preferable to explore the pathogenesis of cryptococcal infection.

2. Intravenous and Intraperitoneal Infection Models

Fatal cryptococcosis in mice can be solidly established by intravenous or intraperitoneal injection of saline suspension of *C. neoformans* cells. The mortality, survival time, and infective rate, as well as severity of clinical signs and histopathological changes, depends on the biological features and inoculum size of the challenging strain of *C. neoformans*, on the one hand, and the natural and/or acquired resistance of the host animal which can be compromised by immunosuppressive treatment, on the other. The work of Rhodes et al.,[338] who studied experimental cryptococcosis in several inbred mouse strains, showed that either decreasing the challenge dose of cryptococci or changing the intravenous route to the intraperitoneal route of inoculation resulted in prolonged survival time, but neither change affected the observed patterns of survival.

Symptomatology and histopathology of experimental cryptococcosis has been described, typically, for infected mice which survived 3 weeks or longer after intravenous challenge with various doses of *C. neoformans* cells. After challenge, mice usually appeared normal until nearly 1 week prior to the time when they developed symptoms of CNS involvement, including distended craniums and locomotor dysfunction. Most mice died several days after the first recognized CNS signs.[339] At necropsy, large numbers of viable cryptococci were recovered from not only the brain, but several other organs including the lungs, liver, spleen, and kidneys.[339-342] Examination of H. E. stained sections from dying or sacrificed animals showed degrees of microscopic lesions in various organs involved. The histopathology of murine cryptococcosis has been characterized by preferential occurrence of either one of the two different types of lesions in organs involved: one was granulomatous and the other cystic.[343-346] In a recent series of intensive studies performed in Miyaji's laboratory, it was clearly demonstrated that such a discrete tissue reaction is relevant to the richness of mononuclear phagocytic cells existing in the tissue which differs from organ to organ; granulomatous lesions predominate in the tissues with plenty of macrophages, such as pulmonary, hepatic, or splenic tissues, whereas cystic lesions occur exclusively in tissues with fewer or a scanty number of macrophages, such as cerebral tissues.[347-350]

Organ distribution and kinetics of distribution of *C. neoformans* in mice administered intravenously was investigated by Duke and Fromtling who used [^{75}Se]-selenomethionine-labeled and unlabeled cryptococci.[339] The injected organisms were rapidly cleared from the blood of mice and then accumulated in the lungs, liver, and kidneys and, to a lesser extent, in the brain within 1 hr. While cryptococci were cleared from the lungs 24 hr after challenge, viable yeasts persisted in the brain, liver, spleen, and kidneys, and additional growth of the organisms occurred only in the brain, lungs, and kidneys. It suggests that *C. neoformans* has a predilection to these three organs.

Heavily or moderately encapsulated yeast cells harvested from cultures of clinical isolates of *C. neoformans* have been generally used as challenge inoculum. So far as described, all such strains belong to serotype A group (perfect state, *F. neoformans*). Yeast cells of *C. neoformans* serotype A seem most likely to be the infectious form and are found abundantly in nature. Many of the *C. neoformans* strains used in the murine model have been reported to achieve 100% mortality in 1 to 4 weeks after infection when mice were challenged with intravenous doses of 10^4 to 10^6 cells. The tendency to believe that heavily or moderately encapsulated *C. neoformans* strains are of greater virulence for mice than thinly encapsulated or nonencapsulated strains of the same species otherwise conferred similar biological properties is generally acceptable. This might be related to the finding that *C. neoformans* cells with small capsules were more easily phagocytized than those with large capsules in the host tissues.[351,352]

The use of a thinly encapsulated, low-virulent pseudohyphal strain of *C. neoformans* has recently been introduced for studies in the pathogenesis of this organism.[352-356] This strain, an isolate from bird excreta, has a low virulence for mice.[354] However, the animals intravenously challenged with this strain developed nonfatal infection exclusively involving the brain, where the viable yeasts persisted for sufficiently long period.[354,357] It was also found by other investigators that the strain could only be isolated from the brain of intraperitoneally infected mice 4 weeks postinfection.[356] Naturally occurring or artificially induced low-virulent or avirulent strains of *C. neoformans* are useful as live vaccine for sensitizing mice or conferring acquired immunity to cryptococcal infection. Reiss and Alture-Werber[358] described a mutant strain which lacks both of the two traits, capsule formation and phenol oxidase activity, characteristic for all virulent *C. neoformans* strains. Their mutant grew well at 5°C and at room temperature and was not reverted to the virulent wild type. Staib and Mishra[354] observed that mice infected with the above-mentioned low-virulent *C. neoformans* strain with predilection to the brain developed severe CNS symptoms with increased mortality when uremia was induced by intramuscular injection of 0.15 mℓ of glycerol in the animals. Fromtling et al.[355] recommended the usefulness of experimental cryptococcosis produced in glycerol-induced acute uremic mice for studies of immune competence in the uremic state.

Several immunosuppressive agents have proven to decrease the survival time or lethal dose of challenge inoculum of cryptococci in mice infected intravenously. These include: whole body X-irradiation,[359] a trichothecen mycotoxin from *Fusarium* diacetoxyscirponol,[360] cyclophosphamide,[339] and antilymphocyte serum.[361] In mice, cyclophosphamide appears to affect host response to *C. neoformans*[362] or impair host resistance to the organism.[339] Monga et al.[340] created B-cell-deficient mice by administration of rabbit antimouse-μ-antiserum to newborn animals for studying the role of antibodies in host defense against experimental cryptococcosis. As a result, they could not find any difference between thus prepared B-cell-deficient mice and the normal mice in the susceptibility to intravenous challenge with *C. neoformans* and the intensity of delayed-type hypersensitivity reaction to the organism.

3. Intracerebral Infection Models

Based on the concept that an appropriate animal model should include established meningitis as the principal site of cryptococcal infection, Graybill and his collaborators developed a murine model of meningitis by direct intracerebral inoculation of cryptococci in order to assess the potential efficacy of several anticryptococcal drugs for oral use.[363-366] They infected mice by injecting 10^2 to 10^4 cryptococci in 0.03 mℓ water delivered from a tuberculin syringe through a 26- to 27-gauge needle inserted 3 to 5 mm into the center of the cranial vault or using a direct puncture in the midline, about 6 mm posterior to the orbit.[363,366] Infected mice exhibited typical progression from normal appearance to ruffled fur, bulging cranium, hunched back, and finally death. Histologically, the mice developed infection characterized by diffuse meningitis and extracerebral dissemination.[363]

4. Intranasal Infection Models

Several investigators prefer the intranasal route of inoculation to produce experimental cryptococcosis because it is considered to be the normal route of infection in humans. In mice, respiratory cryptococcosis has been induced by exposing the animals to soil seeded with *C. neoformans*[367] and, more recently, by intranasal instillation of saline suspension of cryptococci[334,368-370] or by inhalation of aerosolized suspension of the yeasts.[332,335]

In instillation model, lightly anesthetized mice are injected intranasally with 0.005 to 0.05 m*l* of a suspension of inoculum containing approximately 10³ to 10⁴ viable cells of *C. neoformans* delivered onto one nostril with a 20-gauge needle.[334,369,370] Lim et al.[369] followed *C. neoformans* growth in various tissues of mice infected with a relatively small dose of inoculum (10³) by the intranasal instillation route in comparison with that of animals infected intraperitoneally. The majority of mice infected intranasally did not have cryptococci in the lungs immediately after inoculation. Instead, the organisms were found in the head region, but not the brain or mandible. Most likely, after intranasal infection, cryptococci lodged in the nasal passages and either remained there for a period of time without multiplying, or colonized the membranes of the nasal passages before entering the lungs. Entrance to the lungs could have occurred by the organism moving down the respiratory tract and/or by hematogenous or lymphatic spread. In the intraperitoneally infected mice, cryptococci would have to become blood borne to populate the lungs.[369] Ritter and Larsh[334] studied the incidence of mortality and the infective pathway of *C. neoformans* in mice following an intranasal instillation of 10⁴ viable cells. Mice appeared to be more resistant to intranasal route of infection than intraperitoneal route of infection. At the end of 12 weeks, 83% of the mice inoculated intraperitoneally died, whereas the average percentage of death for intranasally inoculated mice was 54%.[334] These investigators also observed that *C. neoformans* was recovered from all organs, including the liver, spleen, kidneys, brain and heart, of the intranasal group of mice at the 4th and 5th week postinfection, although the lungs were most frequently infected. At the 8th week, the lungs appeared as a sole site of infection and after 12 weeks postinfection, 5% of infected mice had cryptococci only in the lungs and brain.[334]

Karaoui et al.[332] induced pulmonary cryptococcosis by using the Henderson apparatus, a modification of that apparatus described by Henderson[371] to expose the mice to aerosols containing the organisms generated by a collison nebulizer. Pulmonary cryptococcosis was not established until day 7 postinfection, probably for the same reason as that observed for the instillation model. Once established, however, it served as the primary source of cryptococci for subsequent dissemination to the liver, the spleen, and finally to the brain. At day 56, *C. neoformans* was recovered from the lungs (100%) and brain (33%), but not cultured from any other organs.[332] This suggests the existence of an alternate line of dissemination to the CNS besides the hematogenous spread as postulated by Wilson.[372] Thus cryptococcosis acquired by the respiratory route should offer a more valid model for the natural infection than cryptococcosis acquired by the artificial route of inoculation.

5. Applications
a. Immunological Studies
Because of the development of useful murine models of systemic cryptococcosis, studies of the immunological responses to *C. neoformans*, especially those involved in the defense against cryptococcal disease, have rapidly progressed in the last few decades. Establishment of several different types of murine models has enabled us to use various approaches to explore the very complicated functioning immune systems. For example, a number of investigators have attempted to stimulate protection of mice against cryptococcosis by immunization with several different types of vaccines, such as killed whole cells or cell fractions[324,373,374] and live cells of virulent or avirulent strains[329,358,369,374-377] administered via various routes, viz., intravenously,[374] intraperitoneally,[369,375] intranasally,[369,376] directly by the pulmonary route,[369,376] intracutaneously,[378] or subcutaneously.[329] The established murine models of cryptococcosis have made it possible to perform extensive studies on the possible role of several immune

responses in the defense against this disease. The voluminous information accumulated suggests a questionable role of antibodies in protection of mice from cryptococcal infection. Although there are few papers which showed a significant increase of survival time of *C. neoformans* infected mice receiving passive transfer of antibodies,[379,380] larger numbers of groups of investigators were unable to protect mice, as well as other experimental animals, from a challenge infection with cryptococci after passive transfer of antibodies.[381-384] On the other hand, evidence for a crucial role of cell-mediated immunity which can be typically expressed as delayed-type hypersensitivity reaction in host-defense against cryptococcosis has been provided by a number of studies with adequately designed experimental systems using murine models[369,370,378,385-387] or guinea pig models.[388] Primary dependence of expression of cell-mediated immunity on T-lymphocytes and mononuclear phagocytes has become increasingly evident and, in murine models, the major role of nonspecifically activated macrophages in resistance to *C. neoformans* infection has firmly been proven.[389-391] This line of immunological study, particularly concerned with natural or acquired immunity to cryptococcosis, has been remarkably accelerated by introduction of certain inbred strains of mice with a selective immunodeficiency of cell-mediated or T-cell-dependent immune function. Congenitally athymic (nude) mice and, less frequently, New Zealand Black (NZB) mice have been successfully used by a number of investigators for exploring the possible role of cell-mediated immunity in the defense against cryptococcal infection. All the published data show an increased susceptibility of nude mice and NZB mice to lethal infections with *C. neoformans* as compared with heterozygous immunocompetent littermate animals and normal CBA mice, respectively.[328,347,348,385-387] These, and a number of comparable results obtained from the experiments in which mice treated with immunosuppressive drug or anti-lymphocyte serum were studied,[324,340,361,362,372,392,393] led us to the consideration that T-lymphocytes play a central role in the defense against cryptococcosis in mice.

Apart from the role of cell-mediated immunity, the importance of humoral factors or, more precisely, complement (C′) systems in defense against cryptococcal infection has been stressed by a large number of workers who performed studies in human and animal models[394-400] since early work on properdin consumption in fatal murine cryptococcosis.[396] In recent years, those types of inbred mouse strains which are genetically deficient in the 5th component of C′ (C5⁻) (e.g., B10.D2/oSn mice) were developed, and they provided a useful and expedient model for studying the possible role of the alternate C′ pathway in the resistance of animals to cryptococcal infection. Rhodes and his co-worker have demonstrated that C5⁻ mice were more highly susceptible to fatal infection with *C. neoformans* than C5-sufficient mice otherwise with the same genotype, and that this relative difference in susceptibility correlated with phenotype of the animals and was a stable, inheritable trait which is probably under the control of a single gene.[338,341] All of evidence tempted Fromtling and Shadomy[401] and Griffin[402] to propose a general scheme of the protective immune system toward *C. neoformans* which suggests that it may depend on a complex interaction of both cellular and humoral immune factors, i.e., stimulation of a sensitized host by cryptococci may trigger a delayed-type hypersensitivity reaction which serves primarily to initiate lymphokine production and activate macrophages to kill phagocytized yeast cells.

b. Evaluation of Antifungal Regimens

Some of the murine models of systemic cryptococcosis are considered to be useful for testing in vivo therapeutic efficacy of antifungal agents. Mice infected with fatal doses of *C. neoformans* intravenously,[342,359,403-407] intraperitoneally,[362,364,368,408,409] intracerebrally,[363-366] or intranasally[368,370] have been employed to evaluate anticryptococcal efficacy of antifungal drugs. Occasionally, experimental models were produced in X-irradiated mice,[359] steroid-treated mice,[370] or nude mice[364,408] with the intention to test in vivo activity of drugs under conditions which mimic those for the compromised human host. In the murine models, decreased incidence of mortality, decreased viable counts cultured from involved tissues, and increased survival time are usually used as criteria for effectiveness of drug activity. Studies with one or more of the murine models have proven a significant protective and/or therapeutic efficacy of the following drugs: flucytosine,[342,359,365,403,406,408] amphotericin B,[342,359,365,368,370,406,408,409] ketoconazole,[362-364,368,406,409] BAY n 7133,[364,407] miconazole,[368] itraconazole,[366] and ICI 153,066,[364] used alone or in appropriate combinations.

C. Other Animal Models with Special Reference to Some Specific Clinical Forms of Cryptococcosis

1. Introduction

Several specific forms of cryptococcosis, including cutaneous, ocular, and gastrointestinal infections, have been successfully produced in mice following challenge with *C. neoformans* via selected routes of inoculation. The rabbit, although a convenient animal for experimental purposes, has been much less frequently used because this species seems to be innately resistant to cryptococcosis. Rabbits tolerated intravenous, intraperitoneal, and intracisternal injection of large inocula of *C. neoformans* without developing persistent or fatal infection.[410] In rabbits, localized infections without accompanying dissemination have been produced in the skin,[411] in the anterior chamber of the eye,[412] and in part of the lungs[413] by direct topical inoculation of cryptococci. More recently, Perfect et al.[414] have developed a new model for cryptococcal meningitis in cortisone-treated rabbits.

Like rabbits, guinea pigs are resistant to cryptococcal infection, whereas rats are more susceptible. All these three species of laboratory animals have not been studied extensively as compared with mice. Only a limited number of papers were published which dealt with experimental cryptococcosis induced, with or without cortisone-treatment, in either of the species. Most of the animal models reported to be established are those of the localized forms of the disease such as cutaneous, pulmonary, ocular, or myocardial cryptococcosis, which have never been reported with mice. Potential applications of these models include studies of the mechanism of susceptibility and resistance to cryptococcosis as well as assessment of antifungals.

2. Cutaneous Infection Models

Experimental cutaneous cryptococcosis has been produced in a variety of laboratory animals, although predominantly in mice.[329,345,415-421] Usually the subcutaneous route of infection has been selected. However, mice or guinea pigs have also been unsuccessfully infected by intradermal injection or by a direct application to the scarified skin with inocula of *C. neoformans*.[420] The general course of the infection in mice was described by several investigators.[329,378,415,418,420] Cutaneous or subcutaneous cryptococcosis was not fatal for the great majority of mice infected, but developed lesions confined to the dermis that did not resolve quickly. Rather, the yeast inoculum survived and proliferated for a month or longer and very few animals died of systemic cryptococcosis. Probably dissemination to the internal organs occurred through hematogenous or lymphatic spread.[329,416,418,420] Conflicting results were obtained by Bergman[345] who observed that a majority of subcutaneously infected mice succumbed to cryptococcosis. Staib and Mishra[354] reported the incidence of death among mice which were inoculated intramuscularly with a high-virulent isolate of *C. neoformans*. Cutaneous infections caused by a low-virulent or avirulent *C. neoformans* strain have often been used to study protective immune responses to subsequent challenge of fatal doses of the organism.[329,377,421]

3. Chronic Meningitis Models

Felton et al.[413] produced localized pulmonary lesions in rabbits by direct inoculation of *C. neoformans* into the lung. They observed a mononuclear inflammatory reaction in the meninges and noncaseating granulomas in the brain, but were unable to demonstrate cryptococci in the CNS by microscopic or culture studies. In more recent years, Perfect et al.[414] developed a new rabbit model of chronic cryptococcal meningitis using cortisone-treated animals. They injected 0.3 m*l* of a suspension of cryptococci into the cisterna magna of each rabbit through a 25-gauge needle, under sedation. After intracisternal challenge, animals developed chronic progressive meningitis that was fatal in 2 to 12 weeks. Incidence and severity of infection was related to cortisone dose, but not to inoculum size. In addition to a similarity to the human disease, this rabbit model has an advantage over murine model in that, because of the larger size of rabbits, repeated aspiration of useful volumes cerebrospinal fluid may be possible. This model was workable for studying the effect of anticryptococcal antibodies on cryptococcal meningitis[422] as well a the therapeutic efficacy and pharmacokinetics of several systemic antifungal agents, such as ketoconazole, amphotericin B, and flucytosine.[423]

4. Pulmonary Cryptococcosis Models

Graybill et al.[424] experimentally induced cryptococcal infection in normal rats and congenitally athymic (nude) rats by inoculating *C. neoformans* cell suspensions intratracheally. They gave the challenge inoculum directly into the right main stem bronchus via a blunt-ended feeding needle inserted through the trachea, which had been surgically exposed, and passed through beyond the carina. After inoculation, immunocompetent rats developed hard nodular lesions in the ipsilateral lung which contained caseous centers with numerous cryptococci. Over the course of a 9-month period, the lesions shrunk and were resolved. In contrast, nude rats developed progressive cryptococcosis with widespread dissemination. These investigators considered that rats may provide a valid model for studying pulmonary cryptococcosis. Gadebusch and Gikas[425] compared the effect of cortisone on pulmonary cryptococcosis experimentally induced in rats and guinea pigs. Cortisone-treatment rendered the disease more severe in rats, and latent infections were manifested in both species of animals.

5. Ocular Cryptococcosis Models

In rabbits, ocular cryptococcal infection has been induced by injecting saline suspensions of *C. neoformans* directly into the anterior chamber of the eye[411] or by the intravenous route.[426] When challenged intravenously, cryptococci were cultured from chorioretinae, as well as from brains and kidneys, of a majority of rabbits 1 to 2 weeks after inoculation, without any ophthalmoscopically observable lesions.[426]

In mice, ocular cryptococcal infections have been produced by a greater number of investigators who injected inocula of *C. neoformans* directly into the eye[427-429] or intravenously.[430-432] Using the low-virulent strain with predilection toward the brain,[354] Staib et al.[357] could induce nonfatal cerebral cryptococcal infection with selective involvement of the eyes. Glycerol-induced uremia increased the incidence of selective involvement of the CNS including the eye.[357]

Blouin and Cello[432] selected the cat as experimental model and infected each animal by the intra-arterial route of inoculation. Under dissociative anesthesia, the left common carotid artery was surgically exposed and 1 m*l* of cryptococcal suspension was injected through a 25-gauge needle. Infected cats developed a progressive, multifocal chorioretinitis comparable to the naturally occurring feline disease. The severity of the ocular disease, the incidence of spread of infection into the fellow eye, and the incident dissemination to the internal organs were shown to be related to the inoculum size. None of the cats died over the 32-day observation period.[432]

6. Myocarditis Model

Nagai et al.[433] developed experimental cryptococcal myocarditis models by inoculating *C. neoformans* intrarenally into rabbits and rats. On the 16th day (or later) postinfection, both species of infected animals developed myocarditis in the heart, with accompanying lesions characterized by focal necrosis and infiltration of small round cells in the myocardium.

7. Gastrointestinal Colonization Model

Green and Bulmer[434] inoculated mice with *C. neoformans* by the oral route through tubing inserted into the stomach. Of the inoculated animals, 24% shed viable cryptococci in feces for several weeks after inoculation and approximately 3% of the animals died of disseminated cryptococcosis 9 to 19 weeks after inoculation. Pre- and posttreatment with cortisone failed to increase susceptibility to *C. neoformans*.[434]

IV. MYCOTIC KERATITIS

A. Introduction

Mycotic keratitis (keratomycosis) is an infection of the cornea by fungi. Although a rare disease previously, it has been occurring with increasing frequency within the past three decades.[435] Both clinical and experimental evidence indicates that the increased incidence of mycotic keratitis is associated with widespread use of corticosteroids, immunosuppressive drugs, and broad-spectrum antibiotics.[436-438]

The normal cornea is resistant to invasion by fungi. It is well known that corneal mycoses, as well as other ocular mycoses, mostly occur after an ocular injury or after a preceding ocular infection or inflammatory disease. Often patiens with such diseases have been treated topically or systemically with antibacterial antibiotics or steroids; these treatments render the cornea more susceptible to fungal invasion. Thus, mycotic keratitis should be considered an opportunistic fungus infection.

The major fungal agents involved in mycotic keratitis are members of the genus *Candida*, particularly *C. albicans* and, less frequently, *C. guilliermondii, C. pseudotropicalis, C. parapsilosis,* and *C. krusei*,[439-441] members of the genus *Aspergillus,* most commonly *A. fumigatus, A. flavus,* and *A. niger;* and *Fusarium solani*.[435,442,443] Besides, a diversity of fungi, which have been considered nonpathogenic or low-pathogenic to humans, have been isolated as causal organisms from cases of mycotic keratitis. They include both yeastlike fungi, such as *Rhodotorula rubra* (*Rhodotorula mucilaginosa*) and *Trichosporon* sp., and filamentous fungi, such as *Curvularia lunata*,[444,445] *Cylindrocarpon tonkinensis, Botryodiploidia theobromae,* and *Phialophora gougerotti*.[446] In addition, each of *Drechslera spinifera*[447] and *Rhizoctonia* sp.[448] was isolated from one case each of mycotic keratitis.

Such increasing incidence of the disease during the last decade, its serious lesions, and the absence of therapeutic antifungal drugs available for ocular applications make it a significant problem in present day medicine. This serious situation has motivated many scientists to develop those animal models of mycotic keratitis which are workable for preclinical evaluation of the antifungal agents, as well as for study of the etiopathology of the disease. With very few exceptions, all the animal models have been produced in rabbits, with or without pretreatment with steroid, following intracorneal inoculation of a variety of fungi.

B. Rabbit Models of Mycotic Keratitis

1. Introduction

The rabbit has been preferentially used to produce the model with a sustained, progressive corneal infection with fungi, probably because rabbit eyes have long been recognized by ophthalmologists as a favorable and easy-to-handle experimental models and also have been shown to be sufficiently susceptible to experimental infections with most of the fungal pathogens of keratitis in humans. Many of the rabbit models of fungal corneal disease thus far developed appear to parallel, partly at least, human disease in terms of clinical manifestations and histopathology. However, the severity and duration of the disease established is influenced by several experimental conditions, such as inoculation techniques and size of infecting inoculum, and, more strongly, by topical or systemic pretreatment of the animal with steroid or other predisposing agents.

2. Infecting Inocula

The infecting organism that has been most often used to obtain experimental mycotic keratitis in rabbits is *C. albicans*, and it is followed by *A. fumigatus* and *F. solani* in this order. Besides these three species of fungi, which predominate as the causal agents of human keratomycosis, stock cultures and clinical isolates of a number of low-pathogenic, saprophytic fungi have been tested for their capacity to induce experimental infections in the rabbits' cornea with or without positive results. The list of such fungi include: *C. krusei*,[441] *R. rubra*,[449] *A. niger*,[450] *A. terreus*,[450,451] *Geotrichum* sp., *Cephalosporium* sp.,[451] *B. theobromae, Pseudallescheria boydii*,[452] and *Rhizoctonia* sp.[448]

C. albicans and other yeast-like fungi — Standard strains from stock cultures and strains isolated from the patient with *Candida* keratitis[441,453] or cutaneous candidiasis[454] have been successfully employed for infecting animals; there may be considerable variances in the pathogenicity to rabbit eye among different strains of *C. albicans*.[455] Following intracorneal inoculation of *R. rubra* isolated from a human deep keratitis, there was no evidence of development of progressive infection in normal or cortisone-treated rabbits.[449] Cultures of yeastlike fungi can be grown well on conventional microbiological solid agars, such as Sabouraud dextrose agar, yeast-extract agar, and Trypticase agar, at 25 to 37°C. Yeast cells are harvested from 2- to 4-day-old cultures, washed off the medium, and finally suspended in sterile saline to prepare the experimental cell suspensions.

Aspergilli, Fusarium and other filamentous fungi — With *A. fumigatus* and other aspergilli, well as *F. solani*, spores are usually harvested from cultures grown on Sabouraud dextrose agar or potato-dextrose agar at 25 to 28°C for 3 to 15 days and suspended in sterile saline, the resultant suspension being employed as the inoculum. The duration of incubation period should be that at which abundant sporulation has occurred. Foster et al.[452] selected an isolate of *F. solani* from a known virulent ulcer which grew well at 37°C for preparing inoculum of this fungus. Such thermotolerant strains would be better at producing progressive infections in animal tissues. Different from most other investigators who used the inoculum of *A. fumigatus* or *F. solani* which consisted mainly of dormant spores, this group of investigators has prepared *A. fumigatus* inoculum from germinated spores of this fungus that can be induced from the dormant spores after 3 to 4 hr incubation. A dematiaceous pigmented fungus *B. theobromae* was grown for several weeks on banana agar until pycnidia were evident.[452] Then the pycnidia were isolated and crushed to release conidia. When sporulation is poor, the mycelial growth obtained from shaking cultures in Sabouraud dextrose broth can be employed to prepare the inoculum as was done by Srivastava et al.[448] with *Rhizoctonia* sp.

3. Methods of Inoculation

The rabbit models of mycotic keratitis have been successfully produced in the animals following intracorneal inoculation of the pathogen suspension under general and/or topical anesthesia. Several different inoculation techniques, such as superficial corneal abrasion (or scarification), trephination (or microtrephination), and inter- (intra-) lamellar injection.

Some groups of investigators have induced *Candida* keratitis in the rabbit eye using the "Cignetti" method, in which the inoculation was performed by making cross-hatched scratches on the anterior part of the corneal stroma with a toothed chalazion currette filled with the *C. albicans* cell suspension.[456-459] Alternatively, the cell suspension in an approximate volume of 1 ml was spread over the abraded areas with a platinum spatule.[460] A sterile scalpel was also used by some others to make superficial scarification of the cornea.[461,462]

In order to establish the rabbit model of mycotic keratitis more effectively, the trephination method has often been employed in which the fungal cell or spore suspension was inoculated into open corneal wounds made by sterile glass trephine.[451,455,463] Oji and his co-workers[463-466] developed a method to perform multiple inoculation in rabbit cornea with microtrephination which could produce a highly reproducible quantitative model of fungal infections of the rabbit corneal stroma. The principle of this method was based on the technique that had been developed and used for measuring antiviral effects in vivo.[467-469] Seventeen corneal wounds (microtrephinations), half corneal thickness, were cut in a regular pattern in each cornea of the rabbits, using a sterile glass trephine 1.5 mm diameter and 100 mm long, under the operating microscope, and the trephine tube loaded with the pathogen suspension was then inserted into each of the trephine sites. This model may enable us to obtain a great deal of data based on multiple microtrephine inoculation sites that would allow reliable statistical analysis in a small number of animals. O'Day et al.[455] also described a model of *C. albicans* infection in the rabbit cornea, using the trephination method. They trephined eight wells, one-half stromal thickness in depth, in the cornea of each eye in a regular pattern with a glass trephine prepared from a hematocrit tube 1 mm in diameter. Each well was then inoculated using a second trephine loaded with the inoculum.

A larger number of the rabbit models of mycotic keratitis have been produced by simple injection of infecting inocula into the corneal stroma. Many investigators have described the infection protocol with this interlamellar injection method which can be generalized as follows: the corneas of anesthetized rabbits are injected intralamellarly with 0.02 to 0.1 ml volumes of the suspensions of fungal cells or spores containing 10^5 to 10^9 viable or particle units per ml by a 25- to 27-gauge needle attached to a tuberculin-type or other type of microsyringe.[441,448-453,470,471,473-478] As compared with the trephination method, the interlamellar inoculation of the inoculum tends to produce a deep and severe infection of the cornea that lasts for longer periods.[475] In the present model, the inoculum may be deposited deep within the corneal stroma and produce a persistent infection, while the epithelium remains undisturbed. Therefore, this interlamellar model would be useful for studying antifungal agents under conditions that more closely parallel human disease.[475]

Ishibashi and his co-workers[454,479-481] have developed a modified method for intracorneal injection. In this method, an intralamellar pocket was first made in the mid-

cornea and then 0.01 ml of the inoculum was injected into the pocket with a 27-gauge needle on a microsyringe. Using this inoculation technique, these investigators have successfully established the reliable model of *Candida* and *Fusarium* keratitis in cortisone-treated rabbits which has proven useful for evaluation of antifungal drugs.[454,480]

4. Effects of Several Experimental Conditions

As is the case with many other experimental infections, the severity of infection in the established models of mycotic keratitis depends on the size of infecting inoculum. Working with the *Candida* keratitis model obtained after the scarification method for inoculation, Segal et al.[461] demonstrated that the rabbits inoculated with 10^4 *C. albicans* cells developed a mild infection within 10 to 14 days, whereas the rabbits inoculated with 10^6 organisms developed, within 5 to 7 days, a severe ocular infection with more severe clinical signs, such as corneal ulcer and hypopyon. A similar effect of the inoculum size was also observed by the same investigators with experimental *A. fumigatus* keratitis; inoculation of 10^4 spores did not produce infection; 10^5 spores yielded variable results; and 10^6 spores, with or without cortisone treatment, resulted in a severe infection.[462] The microtrephination model of *C. albicans* infection also clearly demonstrated the dose-relationship of the pathogens and the host animal, with the lower concentration suspensions of the pathogens producing lower rates of infectivity while the higher concentrations produce higher rates of infectivity.[463]

Since the earlier work of Lay[451] who reported that cortisone applied to traumatized rabbit corneas in the presence of pathogenic fungi resulted in increased incidence of culture-positive mycotic keratitis, the animal models of *C. albicans, A. fumigatus*, or *F. solani* keratitis have been produced more often in rabbits which had been subconjunctivally or topically treated with steroids than in untreated normal animals.[441,448,454,456-458,460,463,472,473,477-480,482-484]

A more direct comparison in the susceptibility to experimental keratomycosis was made by several investigators between the steroid-treated and untreated control rabbits.[450,452,461,462] Foster et al.[452] successfully established corneal infections with or without steroid pretreatment of the rabbits following intralamellar injection of actively germinating conidia from thermotolerant strains of *F. solani*. In this model, although sustained culture-positive ulcers were produced in 35 to 40% of the animals 2 weeks later, pretreatment with subconjunctival steroid was necessary to produce progressive culture-positive ulcers in higher percentage of eyes at 2 and 3 weeks.[452] François and Rysselaere[450] suggested that topical or subconjunctival administration of steroid upon intracorneal inoculation of *F. solani* and aspergilli aggravates the normal lesions by a lowering of the tissue defense reaction and an inhibition of the inflammatory reaction, thus leading to an easier invasion and more pronounced proliferation of mycelia. O'Day et al.[484] reported that topical steroid given alone not only worsened experimental mycotic keratitis but also adversely influenced the efficacy of several antifungal drugs, such as pimaricin, flucytosine, and miconazole, when given in combination.

Several other predisposing agents, such as intravenously administered antilymphocytic serum and whole-body X-irradiation, have been also demonstrated to have a significant potentiating effect on *Candida* keratitis in rabbits.[460,485] Development of experimental mycotic keratitis looks to be enhanced by superimposed bacterial infections. Mahan et al.[478] reported that corneal ulcers produced in normal rabbits or steroid-treated rabbits by intracorneal injection of *A. fumigatus* together with *Staphylococcus aureus* were more severe than those produced after intracorneal injection of *A. fumigatus* alone.

5. Clinical Signs and Histopathology

Infections in the cornea usually begin 24 to 48 hr after intracorneal inoculation of reasonable amounts of suspension of fungal cells or spores and last for varying periods depending on the size of infecting inoculum, predisposing treatment, and/or inoculation technique employed.[441,454-458,463-466,470,471,473,475,479] When the infection is established in the animals, the eyes develop several clinical signs characteristic of keratitis, such as corneal ulcer, hypopyon, and proptosis and, sometimes, result in perforation of cornea.[441,454,462,471,473] During the progressive stage of infection, conjunctival secretions and corneal scrapings are often culture-positive.[441,454,455,471,479]

Histopathologically, the infected eyes usually show a pronounced inflammatory cell infiltration, absence of corneal epithelium and destruction of corneal stroma,[441,471,475] and, moreover, presence of extensive mycelial or mycelial-phase growth of fungi within the stroma.[454,471,475]

The severity of corneal lesions has often been read using several types of scoring systems that were graded according to the area, density, and/or intensity of the lesions.[454,466,486] Such systems may be useful to quantify the effects of antifungal agents.

6. Applications for Assessment of Antifungal Agents

Almost all the rabbit models of mycotic keratitis have been originally developed to provide the reliable and reproducible in vivo system sensitive enough for accurate study of quantitative effects of the antifungal drugs on the eye in reasonable numbers of animals. Various models of corneal infections with *C. albicans*, *A. fumigatus*, and *F. solani* have been employed by many investigators for testing the currently available antifungal drugs and demonstrated that most of them had curative and/or protective effect when administered topically and/or systemically. Such antifungal drugs to which the animal models favorably responded are amphotericin B (intravenous or topical),[455,458,471] pimaricin (natamycin, topical),[455,465,470,471,473] clotrimazole (topical),[463] econazole (topical),[466] miconazole (intravenous or topical),[441,455,462,463,471,479] ketoconazole (oral or topical),[453,454,464,470] and flucytosine (oral or topical).[455,461,475] On the other hand, conflicting results have been reported with some drugs; topical application of miconazole or nystatin to *C. albicans* keratitis models[457] or subconjunctival treatment with pimaricin of *A. fumigatus* keratitis in rabbits[487] has failed to show efficacy.

C. Other Animal Models

There are only a limited number of studies in which experimental mycotic keratitis has been induced in species other than the rabbit. Foster et al.[452] initially used owl monkeys to obtain *A. fumigatus* or *F. solani* keratitis, but they replaced them in later experiments with albino or pigmented rabbits, since it was found that owl monkeys were no more advantageous than rabbits. Burda and Fisher[488] failed to produce experimental *Cephalosporium* keratitis in untreated or cortisone-treated mature rabbits by intracorneal inoculation with a saline spore suspension because of possible inherent antifungal properties of the animal. Thus, these workers abandoned the rabbit as a model, and successfully used mature rats who could establish culture-positive corneal ulcers when the animals were treated either systemically or topically with steroids after intracorneal injection of spores of each of several filamentous fungi including *Cephalosporium* sp., *Fusarium* sp., *A. fumigatus*, and other aspergilli.[488]

V. ZYGOMYCOSIS

A. Introduction

Zygomycosis (mucormycosis) can be defined as a disease caused by various zygomycetes. The synonymous term "phycomycosis" is no longer used because fundamental changes in the classification of fungi have made it obsolete.[489,490] Twelve species of zygomycetes, classified in eight genera, are now listed as well-authenticated causative agents or human zygomycosis:[491] *Absidia corymbifera, Basidiobolus haptosporus, Conidiobolus coronatus, C. incongruus, Cunninghamella bertholletiae, Rhizomucor pusillus, Mucor ramosissimus, Rhizopus microsporus, R. oryzae, R. rhizopodiformis, R. arrhizus,* and *Saksenaea vasiformis.*

In immunocompromised patients, primary pulmonary infection is the most frequent form of mucormycosis. Infection is often acquired through inhalation of spores with subsequent germination and initial proliferation of hyphae which occurs in the lower bronchi and alveoli.[492] However, depending upon the portal of entry, the disease also involves the rhinofacial-cranial area, gastrointestinal tract, skin, and, infrequently, other organ systems.[493] As a result, its clinical forms develop not only as pulmonary infections, but also as cutaneous, subcutaneous, rhinocerebral, and systemic infections. Invasion by the hyphae of the walls and lumens of blood vessels is common and leads to thrombosis and infarction of surrounding tissues. Venous structures may also be involved, in which case tissue hemorrhage is prominent. Hematogenous or lymphatic dissemination of the infection tends to occur if host defense mechanisms are compromised. Individual cases of zygomycosis are identified by demonstration of characteristic mycelial forms of histological sections from isolated tissue. The mycelium of these fungi is nonseptate and is much broader than that of the fungi with a filamentous tissue form. Many of the human pathogenic zygomycetes are also recognized as opportunistic pathogens of domestic animals.[494-496] Various zygomycetous fungi have been implicated in mycotic abortion of cattle.[494]

Experimental studies on zygomycosis have been done with both normal and predisposed laboratory animals challenged with spores of zygomycetes by various routes of inoculation. Mice have been intravenously injected or intranasally instilled with spores of *A. corymbifera (A. ramosa)* and *R. pusillus,* and internal lesions with or without the death of the animals have been reported.[497-510] Intracerebral and intraperitoneal injection of *A. corymbifera* and *R. arrhizus* spores has been reported to be effective in causing mild or severe lesions in the brain and other internal organs.[497-499,511]

The alloxan-diabetic rabbits, intranasally instilled with spores of various zygomycetes, have been employed as the useful animal models of cerebral zygomycosis.[512-517] Intratracheal and other routes of inoculation have also been employed to establish various forms of zygomycosis in diabetic rabbits.[518-521] Experimental infection with zygomycetes have been much less frequently produced in predisposed rhesus monkeys[522-523] and diabetic rats.[524]

B. Murine Models of Zygomycosis
1. Intravenous, Intracerebral, and Intraperitoneal Models
a. Introduction

Murine models of zygomycosis have been produced by inoculation with spores of several zygomycetous fungi, most of which are human pathogens, into normal and compromised mice by the intravenous, intracerebral, or intraperitoneal routes. The systemic type of zygomycosis, where infection was established by the intravenous route, was most frequently used for studying the etiology, immunology, and histopathology of this disease. Since it is accepted that zygomycosis is usually associated with some predisposing condition which has modified the immunological competence of the hosts, the experimental studies of zygomycosis have often been done in mice treated with corticosteroids or some other immunosuppressive agents in order to modify their natural resistance. A number of papers show that corticosteroids are highly effective in predisposing mice to intravenous infections of zygomycetes.[499,506,525,526] However, the effects of such treatments may not closely resemble the situation in natural immunodeficient states. To circumvent this problem, certain strains of mice which naturally develop selective immunodeficiency disorders, such as congenitally athymic nude (nu/nu) mice and New Zealand Black (NZB) mice, have also been employed to establish murine models of systemic zygomycosis.

b. Fungi for Inoculation

Absidia corymbifera has been most frequently used to infect laboratory mice. It was selected because mice are susceptible to both natural and experimental infection with this fungus.[505,511] Several other human pathogens, such as *R. pusillus, R. oryzae* and *R. arrhizus*, were also demonstrated to be pathogenic for normal and cortisone-treated mice when fungal spores were inoculated intravenously.[497,502] Corbel and Eades[499] and Kitz et al.[506] made comparisons of the effects of intravenous inoculation of various strains of *A. corymbifera*, isolated from clinical or veterinary sources and from the environment, without revealing any fundamental difference among them. Kitz et al.[527] reported the potential of a thermotolerant zygomycete *Radiomyces embreei*, which is not known as an etiologic agent in human beings, to cause lethal zygomycosis in mice.

c. Infections in Normal Mice

According to the experimental procedures described by Corbel and Aedes,[499,502] *A. corymbifera* and some other zygomycetes were grown aerobically on malt agar slopes at 37°C until profuse sporulation had occurred, usually between 2 and 4 days after inoculation, and spores harvested in phosphate buffered saline (pH 7.4) by gentle shaking with glass beads. Then the suspensions were decanted, allowed to stand briefly, and the supernatants separated from the beads and debris. The total number of spores per unit volume was determined by direct counting in a cell-counting chamber. Viability was determined by plate counts on malt-extract agar. Volumes of inoculum used were 0.1 ml for intravenous inoculation and 0.2 ml for intraperitoneal and subcutaneous inoculation. Intracerebral inoculations were done by injecting 0.025 ml volume of suspension into the left cerebral hemisphere.

When C3H mice were inoculated with the *A. corymbifera* suspension thus prepared at doses in excess of 10^3 viable units, by the intravenous route, a variable proportion of mice developed lethal zygomycosis of the CNS within 2 to 8 days; the proportion of mice affected was related to the inoculum size, doses of 5×10^7 spores producing lethal infection in 90 to 100% of the mice.[499,501] At necropsy, fungal hyphae, frequently surrounded by infiltrations of mononuclear cells, could be demonstrated in the brain. Lesions were also often present in the kidneys; in other organs they were rare, but the presence of viable fungal spores could be detected by cultural procedures.[499] Similar clinical and histopathological findings were obtained after CBS mice were intravenously inoculated with *R. pusillus* and *R. oryzae*, as well as *A. corymbifera*.[502] Smith[505] also reported that normal mice inoculated intravenously with 10^5 spores of *A. corymbifera* developed brain and kidney infections. The LD_{50} value for 5-week-old mice was estimated to be 10^4 spores. However, mice up to 21-days-old appear to be more susceptible to zygomycosis than older animals.[499]

Intracerebral inoculation of *A. corymbifera* spores was much more effective in establishing infection than inoculation by the intravenous route, and invariably produced lethal infection even with very small doses of spores.[499] The clinical and histological features of the disease were almost identical with those resulting from intravenous inoculation. After intraperitoneal injection of *A. corymbifera* spores, although most animals remained healthy, typical signs of cerebral infection were produced in dead mice and, at necropsy, nonseptate hyphae were present in the brain, kidneys, and liver and also in the peritoneal exudate.[499] Lesions of the internal organs have been reported to be produced in mice which had been injected spores of *R. arrhizus*, but not of *A. corymbifera*, by the intraperitoneal route.[497,511] Metastatic subcutaneous granuloma developed in mice 5 months after sequential intravenous and intracerebral inoculation of *A. corymbifera* spores.[498] Other routes of inoculation such as the subcutaneous, intranasal, and oral administration were not effective in producing any signs of infection in normal mice.[499]

d. Infections in Cortisone-Treated Mice

Treatment of mice with cortisone (acetate) lessened their resistance to infection with *A. corymbifera* spores, which had been challenged by the intravenous or intraperitoneal route, and promoted dissemination of the fungus.[499,506] The susceptibility to this infection was also increased by reticuloendothelial blockade but was not increased by pretreatment with azathioprine, cyclophosphamide, or antithymocyte serum, suggesting that the natural resistance of mice to *A. corymbifera* infection might be dependent upon phagocytic cell function.[499] Following intravenous inoculation with *A. corymbifera* spores, hyphal growth was confined to the brain and kidneys in the normal mice, whereas in cortisone-treated mice, hyphal growth was found in the liver and lungs, as well as the brain and kidneys.[525,526]

e. Infections in Congenitally Immunodeficient Mice

Intravenous infection of athymic nude mice with *A. corymbifera* spores and infection of aging NZB mice with spores of *A. corymbifera*, *R. pusillus* and *R. oryzae* were effective in establishing systemic zygomycosis.[502,504] NZB mice are known to spontaneously develop a selective deficiency of cell-mediated immune function[528] which becomes more severe with increasing age and is related to selective depletion of the thymus-dependent (T) lymphocyte population.[529] Corbel and Eades[502,504] found that there was no difference between aging NZB mice and normal mice or between nude mice and their phenotypically normal littermates in the susceptibility to lethal infections caused by the zygomycetes in terms of mortality, and clinical and histopathological characteristics. The suggestion is that thymus- (or T-lymphocyte-)dependent immune processes do not play an essential role in primary resistance to zygomycosis.

f. Infections in Alloxan-Diabetic Mice

Of the predisposing diseases to zygomycosis in humans, diabetes mellitus has occurred most frequently. Shofield et al.[497] made comparison of the susceptibility of alloxan-induced diabetic mice with that of normal mice to intravenous infection with *R. arrhizus*, but they could not find evidence showing that chronic alloxan-diabetes alters the experimental infection in mice.

2. Intranasal and Intrasinusly Infected Models
a. Introduction

As described above, murine models of zygomycosis have often been produced by employing an intravenous, intracerebral, or intraperitoneal route of inoculation. However, the relevance of the mycotic infections resulting from such a route of inoculation to pulmonary zygomycosis observed in humans is not clear, since the fatal and widely disseminated hyphal zygomycosis reported in the previous studies differs markedly from the pulmonary infections occurring in humans. This raises a necessity to develop a simple and reliable murine model for pulmonary and disseminated zygomycosis in which was employed a route of infection similar to that observed in human pulmonary zygomycosis. Kitz et al.[506] and Waldorf et al.[507] have successfully developed murine models of pulmonary zygomycosis in cortisone-treated mice, resulting in disease that was similar to human zygomycotic infections.

Zygomycosis in humans is rare in the absence of preexisting disease states known to affect immune competence.[530] One such underlying disease is diabetes mellitus, and its specific association with zygomycosis is well documented.[531] To assess the influence of diabetes in predisposition to pulmonary and cerebral zygomycotic infections, murine models of diabetic zygomycosis have also been developed by Waldorf et al.[509,510]

b. Infections in Cortisone-Treated Mice

Murine models of pulmonary zygomycosis have been established in cortisone-treated mice in intranasal instillation of spores of *A. corymbifera* or *R. pusillus*.[507,508] Spores were harvested from 4-day-old malt-extract agar cultures grown in Roux bottles at 37°C and suspended in aqueous 0.01% Tween 80 solution.[507] Spores thus suspended were administered intranasally using the method of Levine et al.[532] that was originally developed for establishing pulmonary murine coccidioidomycosis. Cortisone acetate, in saline with 0.01% Tween 80, was administered subcutaneously just before the inoculation of spore suspensions.[499,507] Some groups of investigators demonstrated that intranasal inoculation of cortisone-treated mice with the zygomycetous spores resulted in a fatal zygomycotic infection which presents pathology closely resembling pulmonary zygomycosis observed in humans, extensive tissue necrosis in the vicinity of hyphae and a suppurative host response, and that in such infected mice, subsequent systemic fungal dissemination occurred by means of blood-borne thrombi as is again observed in zygomycosis in humans. The 50% infectious doses were 2.4×10^4 CFU for lung infections and 2.7×10^5 CFU for brain infections.[507] Treatment of normal mice with corticosteroids appears to be prerequisite to establish murine models of pulmonary zygomycosis. In uncortisone-treated mice inoculated intranasally with much larger doses of zygomycetous spores, no mortality, infection, or clinical signs of disease were seen, although spores introduced, then localized in the lung, were capable of seeding other tissues.[499,506,507] The absence of germinated *R. pusillus* spores in inoculated but uncortisone-treated mice may be due to a reversible inhibition of spore germination rather than destruction of spores by the host.[508]

c. Infections in Streptozotocin-Diabetic Mice

Diabetes mellitus, especially with ketoacidosis, has been the most commonly recognized underlying disease associated with pulmonary and rhinocerebral forms of zygomycosis.[530] Murine models of these species forms of zygomycosis have been developed by Waldorf et al.[509,510] in diabetic mice. Diabetes was induced by streptozotocin in 4- to 6-week-old, pathogen-free white mice.[509] Streptozotocin in a dose of 250 mg/kg was injected intraperitoneally into mice, in which a mild ketoacidotic diabetes developed within 7 days. Spores of *R. oryzae* were administered intranasally using the inoculation technique described above[507] into the mice 7 days following injection of streptozotocin. The inoculated diabetic mice developed fatal infection with histopathology resembling pulmonary zygomycosis observed in humans. The mean lethal dose was 1.2×10^4 CFU, and 70% of the deaths occurred between days 1 and 4 after inoculation.[509] Experimental cerebral zygomycosis was established when *R. oryzae* spores (10^6 in 0.05 mℓ) were inoculated directly into the ethmoid sinus via the anterior nasal cavity of diabetic mice. Ninety percent of diabetic mice died after inoculation with 10^6 spores. Although intrasinus inoculations caused minimal direct trauma and lower levels of brain inocula than the intranasal route, progressive cerebral infections were induced in diabetic mice.[509,510]

C. Experimental Zygomycosis in Rabbits
1. Introduction

The alloxan-diabetic rabbit has been extensively used for establishing several different forms of zygomycosis. One of the merits of this model is the similarity with a typical form of the disease in compromised human hosts (diabetes). Experimental cerebral zygomycosis, established following intranasal instillation with spores of zygomycetes into diabetic rabbits, has been frequently used to study the etiopathology of zygomycosis and the pathogenicity of various zygomycetes.[512-516] This route of inoculation was also used to infect granulocytopenic rabbits with zygomycetes.[517] A more specific form of pulmonary zygomycosis in diabetic rabbits has been established by employing the intratracheal route of inoculation.[518] Local inoculation of spores of zygomycetes produced cutaneous, subcutaneous, or orbital infections depending on the site of injection.[519-521]

2. Intranasal Models

Cerebral and pulmonary zygomycosis has been established following the intranasal instillation of spores of various species of zygomycetes into ketotic rabbits with alloxan-induced diabetes.[512-517] Spores harvested from 4- to 7-day-old cultures on potato-dextrose agar were suspended in sterile 0.15 M saline, and the suspensions were administered to the rabbits by nasal instillation 48 hr after alloxan treatment.[515,516] For intranasal instillation, 1 mℓ amounts of the spore suspensions (10^7 spores/mℓ) were dropped into nares of the rabbits while they were restrained at a decline of about 30°. With this method of inoculation, Reinhardt and his co-workers[515,516] demonstrated that a number of thermotolerant species of *Rhizopus (R. arrhizus, R. chinensis, R. microsporus, R. oligosporus, R. oryzae,* and *R. rhizopodiformis), A. corymbifera,* and *R. pusillus* cause cerebral zygomycosis in ketotic diabetic rabbits and that most isolates of *Rhizopus* are consistently more pathogenic than isolates of the latter two species. These results support evidence provided earlier by Bauer et al.[512,513] and Kaplan et al.[514]

Intranasal infection with *R. oryzae* was also produced in rabbits with sustained, severe leukopenia and granulocytopenia induced by repeated injections of nitrogen mustard.[517] Initially, these animals developed extensive fungus lesions at the site of inoculation which later became granulomatous and tended to heal. Therefore, the behavior of the infection in this model looks greatly different from the unchecked progression of zygomycosis in the animal with acute alloxan diabetes.

3. Intratracheal Models

Although the above-mentioned procedure is effective in establishing cerebral infections, its value for the production of pulmonary zygomycosis is limited. Instead, intratracheal inoculation of spores into the lungs would be a better method for establishing pulmonary zygomycosis. Elder et al.[518] produced the pulmonary form of zygomycotic infection by intratracheal inoculation of the suspensions of *R. arrhizus* spores into rabbits in the acute toxic phase of alloxan diabetes. For inoculation, a rabbit was anesthetized with ether, and the trachea was surgically exposed. During the injection of the spore suspensions, the rabbit was held on the right side at such an angle as to allow the material to run into the right main-stem bronchus. The infection resulted in great proliferation of the hyphal form of the fungus in the bronchi and lungs.[518]

4. Intradermal, Subcutaneous, and Orbital Models

Sheldon and Bauer[519,520] demonstrated that intradermal and subcutaneous injection of *R. oryzae* spores into rabbits with acute alloxan diabetes and acidosis well-established the cutaneous type and the subcutaneous granulomatous types of infection, respectively, at the site of inoculation. Development of a more specific type of zygomycosis localized in the orbital region in alloxan-diabetic rabbits was reported by Mahajan et al.[521] When *R. oryzae* spores were inoculated into the orbit of normal or cortisone-treated rabbits, no microbiological or pathological evidence of infection was produced. However, orbital infection, as evidenced by proliferating fungi as well as development of necrosis and granulomatous reaction in the retro-orbital tissue, was established in cortisone-treated alloxan-diabetic rabbits.[521] Although in the infected animals fungal growth invasively involved sclera, choroid, and episclera, the infection remained clinically unmanifested.

D. Other Animal Models

Rhesus monkeys and rats have been less frequently used for producing experimental infections with zygomycetes. Disseminated type of lethal infection was established in prednisolone-treated rhesus monkeys which had been injected intravenously with 5 to 9×10^6 spores of *R. rhizopodiformis*.[522] The lesions of the disease in the infected monkeys were fulminated in the kidneys and gastric mucosa, and minimial and focal in the other viscera. Rhino-orbital type of zygomycosis was also produced in rhesus monkeys.[523] Experimental cutaneous zygomycosis of normal and alloxan-induced diabetic rats were produced and used by Sheldon and Bauer[524] to study the role of the tissue mast cells or both groups of animals in relation to the acute inflammatory reaction to this experimental infection.

REFERENCES

1. Raper, K. B. and Fennell, D. I., *The Genus Aspergillus*, Williams & Wilkins, Baltimore, 1965.
2. Young, R. C., Jennings, A., and Bennett, J. E., Species identification of invasive aspergillosis in man, *Am. J. Clin. Pathol.*, 58, 554, 1972.
3. Emmons, C. W., Binford, C. H., and Utz, J. P., *Medical Mycology*, 2nd ed., Lea & Febiger, Philadelphia, 1970.
4. Young, R. C., Bennett, J. E., Vogel, C. L., Carbone, P. P., and DeVita, V. T., Aspergillosis. The spectrum of the disease in 98 patients, *Medicine*, 49, 147, 1970.
5. Meyer, R. D., Young, L. S., Armstrong, D., and Yu, B., Aspergillosis complicating neoplastic disease, *Am. J. Med.*, 54, 6, 1973.
6. Bach, M. C., Adler, J. L., Breman, J., Peng, F., Sahyoun, A., Schlesinger, R. M., Madras, P., and Monaco, A. P., Influence of rejection therapy on fungal and nocardial infections in renal transplant patients, *Lancet*, I, 180, 1973.
7. Burton, J. R., Zachery, J. B., Bessin, R., Rathbun, H. K., Greenough, W. B., III, Sterioff, S., Wright, J. R., Slavin, R. E., and Williams, G. M., Aspergillosis in four renal transplant recipients, *Ann. Intern. Med.*, 77, 383, 1972.
8. Lazarus, G. M. and Neu, H. C., Agents responsible for infection in chronic granulomatous disease of childhood, *J. Pediatr.*, 86, 415, 1975.
9. Baker, R. D., The pathologic anatomy of mycoses, in *Handbuch der speziellen pathologischen Anatomie und Histologie*, Vol. 5 (Part 3), Uehlinger, E., Ed., Springer-Verlag, Basel, 1971.
10. Aslam, P. A., Eastridge, C. E., and Hugh, F. A., Aspergillosis of the lung — an eighteen-year experience, *Chest*, 59, 28, 1971.
11. Ovie, N. G. M., deVries, G. A., and Kikstra, A., Growth of *Aspergillus* in the human lung. Aspergillus and aspergillosis, *Am. Rev. Resp. Dis.*, 82, 649, 1960.
12. Hoehne, J. H., Reed, C. E., and Dickie, H. A., Allergic bronchopulmonary aspergillosis is not rare. With a note on preparation of antigen for immunologic tests, *Chest*, 63, 177, 1973.
13. Jordan, C., Bierman, C. W., and Van Arsdel, P. O., Allergic bronchopulmonary aspergillosis, *Arch. Intern. Med.*, 128, 576, 1971.
14. Ainsworth, G. E. and Austwick, P. K. C., Fungal Diseases of Animals. Commonwealth Agricultural Bureaux, Farnham Royal, Bucks, England, 1959, 1.
15. Chute, H. L., Fungal infections, in *Diseases of Poultry*, 6th ed., Hofstad, B. W., Calneck, C. F., Helmbutt, W. M., et al., Eds., Iowa State University Press, Ames, 1972, 448.
16. Eggert, M. J. and Rombert, P. F., Pulmonary aspergillosis in a calf, *J. Am. Vet. Med. Assoc.*, 137, 595, 1960.

17. Griffin, R. M., Pulmonary aspergillosis in the calf, *Vet. Rec.*, 84, 109, 1969.

18. Long, J. R. and Mitchell, L., Pulmonary aspergillosis in a mare, *Can. Vet. J.*, 12, 16, 1971.

19. Pakes, S. P., New, A. E., and Benbrook, S. C., Pulmonary aspergillosis in a cat, *J. Am. Vet. Med. Assoc.*, 151, 950, 1967.

20. Auswick, P. K. C., Gitter, M., and Watkins, C. V., Pulmonary aspergillosis in lambs, *Vet. Rec.*, 72, 19, 1960.

21. Vitovec, J., Vladik, P., and Fragner, P., Morphologie der Lungenveränderungen bei aspergillose des Rehwildes, *Mykosen*, 15, 189, 1972.

22. Blount, W. P., Rabbit ailments, in *Fur and Feathers*, Idle, Bradford, U.K., 1957, 118.

23. Cohrs, P., Jaffe, R., and Meesen, H., Eds., *Pathologie der Laboratoriumstiere*, Vol. 2, Springer-Verlag, Berlin, 69.

24. Dumas, J., *Les Animaux de Laboratoire*, Editions Medicales, Flammarion, Paris, 1953, 187.

25. Lesbouyries, G., *Pathologie du Lapin*, Societe Anonyme d'Editions Medicales et Scientifiques, Libraire Maloine, Paris, 1963, 245.

26. Seifried, O., *Die Krankheiten des Kaninchens*, Springer-Verlag, Berlin, 1937, 116.

27. Patton, N. M., Cutaneous and pulmonary aspergillosis in rabbits, *Lab. Animal Sci.*, 25, 347, 1975.

28. Turner, K. J., Hackshaw, R., Papadimitriou, J., Wetherall, J. D., and Perrott, J., Experimental aspergillosis in rats infected via intraperitoneal and subcutaneous routes, *Immunology*, 29, 55, 1975.

29. Sideransky, H. and Friedman, L., The effects of cortisone and antibiotic agents on experimental pulmonary aspergillosis, *Am. J. Pathol.*, 35, 169, 1959.

30. Sideransky, H., Verney, E., and Beede, H., Experimental pulmonary aspergillosis, *Arch. Pathol.*, 79, 299, 1965.

31. Rippon, J. W. and Anderson, D. N., Experimental mycosis in immunosuppressed rabbits. II. Acute and chronic aspergillosis, *Mycopathologia*, 64, 97, 1979.

32. Corbel, M. J. and Eades, S. M., Examination of the effect of age and acquired immunity on the susceptibility of mice to infection with Aspergillus fumigatus, *Mycopathologia*, 60, 79, 1977.

33. Turner, K. J., Hackshaw, R., Papadimitriou, J., and Perrott, J., The pathogenesis of experimental pulmonary aspergillosis in normal and cortisone treated rats, *J. Pathol.*, 118, 65, 1976.

34. Sideransky, H. and Verney, E., Experimental aspergillosis, *Lab. Invest.*, 11, 1172, 1962.

35. Ford, S. and Friedman, L., Experimental study of the pathogenicity of *Aspergilli* for mice, *J. Bacteriol.*, 94, 928, 1967.

36. Sideransky, H., Epstein, S. M., Verney, E., and Horowitz, C., Experimental visceral aspergillosis, *Am. J. Pathol.*, 69, 55, 1972.

37. Epstein, S. M., Verney, E., Miale, T. D., and Sideransky, H., Studies on the pathogenesis of experimental pulmonary aspergillosis, *Am. J. Pathol.*, 51, 769, 1967.

38. Merkow, L. L., Epstein, S. M., Sideransky, H., Verney, E., and Pardo, M., The pathogenesis of experimental pulmonary aspergillosis, *Am. J. Pathol.*, 62, 57, 1970.

39. Kish, A. L., Rosenberg, P., Maydew, R., and Southard, L., Studies of the pathogenesis of immunosuppression-induced exogenous and reactivation-type murine aspergillosis, *Clin. Res.*, 25(Abstr.), 156A, 1977.

40. White, L. O., Germination of aspergillus fumigatus conidia in the lung of normal and cortisone-treated mice, *Sabouraudia*, 15, 37, 1977.

41. Schaffner, A., Douglas, H., and Braude, A., Selective protection against conidia by mononuclear and against mycelia by polymorphonuclear phagocytes in resistance to Aspergillus, *J. Clin. Invest.*, 69, 617, 1982.

42. Bhatia, V. N. and Mohapatra, L. N., Experimental aspergillosis in mice. I. Pathogenic potential of *Aspergillus fumigatus, Aspergillus flavus* and *Aspergillus niger, Mykosen*, 12, 615, 1969.

43. Bhatia, V. N. and Mohaptra, L. N., Experimental aspergillosis in mice. II. Enhanced susceptibility of the cortisone treated mice to infection with *Aspergillus fumigatus, Aspergillus flavus* and *Aspergillus niger, Mykosen*, 13, 105, 1970.

44. Sandhu, D., Sandhu, R. S., Damodaran, V. N., and Randhawa, H. S., Effect of cortisone on bronchopulmonary aspergillosis in mice exposed to spores of various *Aspergillus* species, *Sabouraudia*, 8, 32, 1970.

45. Pore, R. S. and Larsh, H. W., Experimental pathology of *Aspergillus terreus-flavipes* group species, *Sabouraudia*, 6, 89, 1968.

46. Pore, R. S. and Larsh, H. W., Aleuriospore formation in four related *Aspergillus* species, *Mycologia*, 59, 318, 1967.

47. Scholer, H. J., Experimentelle Aspergillose der Maus (*Aspergillus fumigatus*) und ihre chemotherapeutische Beeinflussung, *Schweiz. Zeitschr. Pathol. Bakteriol.*, 22, 564, 1959.

48. Smith, G. R., Experimental aspergillosis in mice: aspects of resistance, *J. Hyg. Camb.*, 70, 741, 1972.

49. Smith, G. R., *Aspergillus fumigatus*: a possible relationship between spore size and virulence for mice, *J. Gen. Microbiol.*, 102, 413, 1977.

50. Graybill, J. R., Kaster, S. R., and Drutz, D. J., Treatment of experimental murine aspergillosis with BAY n 7133, *J. Infect. Dis.*, 148, 898, 1983.
51. Rippon, J. W., Anderson, D. N., and Soo Hoo, M., Aspergillosis: comparative virulence, metabolic rate, growth rate and ubiquinone content of soil and human isolates of *Aspergillus terreus, Sabouraudia*, 12, 157, 1971.
52. Purnell, D. M., Effects of specific genotypic alterations on the virulence of *Aspergillus nidulans* for mice, *Microb. Genet. Bull.*, 33, 14, 1971.
53. Purnell, D. M. and Martin, G. M., *Aspergillus nidulans*: association of certain alkaline phosphatase mutations with decreased virulence for mice, *J. Infect. Dis.*, 123, 305, 1971.
54. Purnell, D. M. and Martin, G. M., Heterozygous diploid strains of *Aspergillus nidulans*: enhanced virulence for mice in comparison to a prototrophic haploid strain, *Mycopathol. Mycol. Appl.*, 49, 307, 1973.
55. Purnell, D. M. and Martin, G. M., A morphologic mutation in *Aspergillus nidulans* associated with increased virulence for mice, *Mycopathol. Mycol. Appl.*, 51, 75, 1973.
56. Purnell, D. M., The effects of specific auxotrophic mutations on the virulence of *Aspergillus nidulans* for mice, *Mycopathol. Mycol. Appl.*, 50, 195, 1973.
57. Purnell, D. M., The histopathologic response of mice to *Aspergillus nidulans*: comparison between genetically defined haploid and diploid strains of different virulence, *Sabouraudia*, 12, 95, 1974.
58. Lehmann, P. F. and White, L. O., Chitin assay used to demonstrate renal localization and cortisone-enhanced growth of *Aspergillus fumigatus* mycelium in mice, *Infect. Immun.*, 12, 987, 1975.
59. Lehman, P. F. and White, L. O., Acquired immunity to *Aspergillus fumigatus, Infect. Immun.*, 13, 1296, 1976.
60. Corbel, M. J. and Eades, S. M., The relative suscpetibility of New Zealand Black and CBA mice to infection with opportunistic fungal pathogens, *Sabouraudia*, 14, 17, 1976.
61. Shiraishi, A., Studies on the host defence mechanisms against *Aspergillus* infection, *Chiba Med. J.*, 54, 297, 1978.
62. Purnell, D. M., Enhancement of tissue invasion in murine aspergillosis by systemic administration of suspensions of killed *Corynebacterium parvum, Am. J. Pathol.*, 83, 547, 1976.
63. Sheldon, W. H. and Bauer, H., The role of predisposing factors in experimental fungus infections, *Lab. Invest.*, 11, 1184, 1962.
64. Mankowski, Z. T. and Littleton, B. J., Action of cortisone and ACTH on experimental fungus infections, *Antibiot. Chemother.*, 4, 253, 1954.
65. Piggot, W. R. and Emmons, C. W., Device for inhalation exposure of animals to spores, *Proc. Soc. Exp. Biol. Med.*, 103, 805, 1960.
66. Merkow, L. M., Pardo, M., Epstein, S. M., Verney, E., and Sidransky, H., Lysosomal stability during phagocytosis of Aspergillus flavus spores by alveolar macrophages of cortisone-treated mice, *Science*, 160, 79, 1968.
67. Epstein, S. M., Miale, T. D., Moossy, J., Verney, E., and Sideransky, H., Experimental intracranial aspergillosis, *J. Neuropathol. Exp. Neurol.*, 27, 473, 1968.
68. Graybill, J. R. and Kaster, S. R., Experimental murine aspergillosis. Comparison of amphotericin B and a new polyene antifungal drug, SCH 28191, *Am. Rev. Respir. Dis.*, 129, 292, 1984.
69. Schaffner, A. and Frick, P. G., The effect of ketoconazole on amphotericin B in a model of disseminated aspergillosis, *J. Infect. Dis.*, 151, 902, 1985.
70. Graybill, J. R. and Ahrens, J., Itraconazole treatment of murine aspergillosis, *Sabouraudia*, 23, 219, 1985.
71. Van Cutsem, J., Van Gerven, F., Van de Ven, M.-A., Borgers, M., and Janssen, P. A. J., Itraconazole, a new triazole that is orally active in aspergillosis, *Antimicrob. Agents Chemother.*, 26, 527, 1984.
72. Schar, G., Kayser, F. H., and Dupont, M. C., Antimicrobial activity of econazole and miconazole in vitro and in experimental candidiasis and aspergillosis, *Chemotherapy*, 22, 211, 1976.
73. Plempel, M., Antimycotic activity of BAY N 7133 in animal experiments, *J. Antimicrob. Chemother.*, 13, 447, 1984.
74. Polak, A., Oxiconazole, a new imidazole derivative. Evaluation of antifungal activity in vitro and in vivo, *Arzneim. Forsch./Drug Res.*, 32(I), 17, 1982.
75. Polak, A., Scholer, H. J., and Wall, M., Combination therapy of experimental candidiasis, cryptococcosis and aspergillosis in mice, *Chemotherapy*, 28, 461, 1982.
76. Henrici, A. T., An endotoxin from *Aspergillus fumigatus, J. Immunol.*, 36, 319, 1939.
77. Kurup, V. P. and Sheth, N. K., Experimental aspergillosis in rabbits, *Comp. Immunol. Microbiol. Infect. Dis.*, 4, 161, 1981.
78. Kurup, V. P., Interaction of *Aspergillus fumigatus* spores and pulmonary alveolar macrophages of rabbits, *Immunobiology*, 166, 53, 1984.

79. Hotchi, M., A histopathological study on experimental pulmonary aspergillosis in sensitized animals, *Jpn. J. Med. Mycol.*, 24, 159, 1983.
80. Morin, O., Nomballais, M. F., and Vermeil, C., Experimentally induced aspergillosis in the rabbit, *Mycopathologia*, 54, 63, 1974.
81. Morin, O., Nomballais, M. F., and Vermeil, C., Aspergillose experimentale du lapin. Reponses immunologiques et anatomo-pathologiques a un envahissement pulmonaire unique et massif de spores vivantes d'*Aspergillus fumigatus*. Correlations anatomo-serologiques. Problemes poses par les infestations pulmonaires aspergillaires fugaces, *Mycopathol. Mycol. Appl.*, 54, 63, 1974.
82. Eskenasy, A. and Molan, M., Experimental pulmonary aspergillosis in sensitized rabbits, *Rev. Roum. M. E. P.*, 23, 207, 1977.
83. Weiner, M. H. and Coats-Stephen, M., Immunodiagnosis of systemic aspergillosis. I. Antigenemia detected by radioimmunoassay in experimental infection, *J. Lab. Clin. Med.*, 93, 111, 1979.
84. Andrews, C. P. and Weiner, M. H., Immunodiagnosis of invasive pulmonary aspergillosis in rabbits. Fungal antigen detected by radioimmunoassay in bronchoalveolar lavage fluid, *Am. Rev. Respir. Dis.*, 124, 60, 1981.
85. Damodaran, V. N. and Chakravarty, S. C., Mechanism of production of Candida lesions in rabbits, *J. Med. Microbiol.*, 6, 287, 1973.
86. Edwards, J. E., Jr., Montgomerie, J. Z., Foos, R. Y., Shaw, V. K., and Guze, L. B., Experimental hematogenous endophthalmitis caused by *Candida albicans*, *J. Infect. Dis.*, 131, 649, 1975.
87. Fujita, N. K., Henderson, D. K., Hockey, L. J., Guze, L. B., and Edwards, J. E., Jr., Comparative ocular pathogenicity of *Cryptococcus neoformans*, *Candida glabrata*, and *Aspergillus fumigatus* in the rabbit, *Invest. Ophthalmol. Vis. Sci.*, 22, 410, 1982.
88. Ellison, A. C., Intravitreal effects of pimaricin in experimental fungal endophthalmitis, *Am. J. Pathol.*, 81, 157, 1976.
89. Segal, E., Romano, A., and Barishak, Y. R., Miconazole activity in experimental Aspergillus ocular infections, *Ophthal. Res.*, 13, 12, 1981.
90. Carrizosa, J., Kohn, C., and Levison, M. E., Experimental aspergillus endocarditis in rabbits, *J. Lab. Clin. Med.*, 86, 746, 1975.
91. Tanphaichitra, D., Ries, K., and Levison, M. E., Susceptibility to *Streptococcus viridans* endocarditis in rabbits with intracardiac pacemaker electrodes or polyethylene tubing, *J. Lab. Clin. Med.*, 84, 726, 1974.
92. Turner, K. J., Papadimitriou, J., Hackshaw, R., and Wetherall, J. D., Experimental aspergillosis in normal rats infected intravenously, *J. Reticuloendothel. Soc.*, 17, 300,
93. Slavin, R. G., Fischer, V. W., Levine, E. A., Tsai, C. C., and Winzenburger, P., A primate model of allergic bronchopulmonary aspergillosis, *Int. Archs. Allergy Appl. Immun.*, 56, 325, 1978.
94. Slavin, R. G., Fischer, V. W., Hutcheson, P. S., and Tsai, C. C., Skin tests in a primate model of allergic bronchopulmonary aspergillosis, *Int. Archs. Allergy Appl. Immun.*, 65, 241, 1981.
95. Bendidixen, H. C. and Plum, N., Schimmelpilze (*Aspergillus fumigatus* und *Absidia ramosa*) als Abortusversuche beim Rinde, *Acta Pathol. Microbiol. Scand.*, 6, 252, 1929.
96. Hillman, R. V. and McEntee, K., Experimental studies on bovine mycotic placentitis, *Cornell Vet.*, 59, 289, 1969.
97. Hill, M. W. M., Whiteman, C. E., Benjamin, M. M., and Ball, L., Pathogenesis of experimental bovine mycotic placentitis produced by *Aspergillus fumigatus*, *Vet. Pathol.*, 8, 175, 1971.
98. Whiteman, C. E., Benjamin, M. M., Ball, L., and Hill, M. W. M., Bovine aspergillosis produced by the inoculation of conidiospores of *Aspergillus fumigatus* into a mesenteric or jugular vein, *Vet. Pathol.*, 9, 408, 1972.
99. Cysewski, S. J. and Pier, A. C., Mycotic abortion in ewes produced by *Aspergillus fumigatus*: pathologic changes, *Am. J. Vet. Res.*, 29, 1135, 1968.
100. Pier, A. C., Cysewski, S. J., and Richard, J. L., Mycotic abortion in ewes produced by *Aspergillus fumigatus*: intravascular and intra-uterine inoculation, *Am. J. Vet. Res.*, 33, 349, 1972.
101. Corbel, M. J., Pepin, G. A., and Millar, P. G., The serological response to *Aspergillus fumigatus* in experimental mycotic abortion in sheep, *J. Med. Microbiol.*, 6, 539, 1973.
102. Day, C. A. and Corbel, M. J., Haematological changes associated with *Aspergillus fumigatus* infection in experimental mycotic abortion of sheep, *Br. J. Exp. Pathol.*, 55, 352, 1974.
103. Thurston, J. R., Cysewski, S. J., Pier, A. C., and Richard, J. L., Precipitins in serums from sheep infected with *Aspergillus fumigatus*, *Am. J. Vet. Res.*, 33, 929, 1972.
104. Taylor, J. J. and Burroughs, E. J., Experimental avian aspergillosis, *Mycopathol. Mycol. Appl.*, 51, 131, 1973.
105. Van Cutsem, J., Antifungal activity of enilconazole on experimental aspergillosis in chickens, *Avian Dis.*, 27, 36, 1983.
106. O'Meara, D. C. and Chute, H. L., Aspergillosis experimentally produced in hatching chicks, *Avian Dis.*, 3, 404, 1959.

107. Asakura, S., Nakagawa, S., Masui, M., and Yasuda, J., Immunological studies of aspergillosis in birds, *Mycopathologia*, 18, 249, 1962.
108. Kong, Y. C. M. and Levine, H. B., Experimentally induced immunity in the mycoses, *Bacteriol. Rev.*, 31, 35, 1967.
109. Emmons, C. W., Binford, C. H., Utz, J. P., and Kwon-Chung, K. J., *Medical Mycology*, 3rd ed., Lea & Febiger, Philadelphia, 1977.
110. Odds, F. C., *Candida and Candiosis*, Leicester/University Park Press, Baltimore, 1979.
111. Taschdjian, C. L., Seelig, M. S., and Kozinn, P. J., Serological diagnosis of candidial infections, *CRC Crit. Rev. Clin. Lab. Sci.*, 4, 19, 1973.
112. Krause, W., Matheis, H., and Wulf, L., Fungaemia and funguria after oral administration of Candida albicans, *Lancet*, 1, 598, 1969.
113. Meyerowitz, R. L., Pazin, G. J., and Allen, C. M., Disseminated candidiasis: changes in incidence, underlying diseases and pathology, *Am. J. Clin. Pathol.*, 68, 29, 1977.
114. Louria, D. B., Stiff, D. P., and Bennett, B., Disseminated moniliasis in the adult, *Medicine (Baltimore)*, 41, 307, 1962.
115. Rippon, J. W., *Medical Mycology: The Pathologenic Fungi and the Pathogenic Actinomycetes*, W. B. Saunders, Philadelphia, 1974.
116. Eras, P. L., Goldstein, M. J., and Sherlock, P., Candida infection of the gastrointestinal tract, *Medicine*, (Baltimore), 51, 367, 1972.
117. Parker, J. C., Jr., McCloskey, J. J., and Lee, R. S., The emergence of candidiosis: the dominant postmortem cerebral mycosis, *Am. J. Clin. Pathol.*, 70, 31, 1978.
118. Rayner, C. R. W., Disseminated candidiasis in the severely burned patient, *Plast. Reconstr. Surg.*, 51, 461, 1973.
119. Cho, S. Y. and Choi, H. Y., Opportunistic fungal infection among cancer patients: a ten year autopsy study, *Am. J. Clin. Pathol.*, 72, 617, 1979.
120. Remington, J. S. and Anderson, S. E., Jr., Pneumocystis and fungal infection in patients with malignancies, *J. Radiat. Oncol. Biol. Phys.*, 1, 313, 1976.
121. Zimmerman, L. E., Fatal fungus infections complicating other diseases, *Am. J. Clin. Pathol.*, 25, 46, 1955.
122. Ray, T. L., Fungal infections in the immunocompromised host, *Med. Clin. N. Am.*, 64, 955, 1980.
123. Gaines, J. D. and Remington, J. S., Disseminated candidiasis in the surgical patient, *Surgery*, 72, 730, 1972.
124. Montgomerie, J. Z. and Edwards, J. E., Jr., Association of infection due to *Candida albicans* with intravenous hyperalimentation, *J. Infect. Dis.*, 137, 197, 1978.
125. Thienpont, D., Van Cutsem, J., and Borgers, M., Ketoconazole in experimental candidosis, *Rev. Infect. Dis.*, 2, 570, 1967.
126. Fiennes, R., *Zoonoses of Primates: the epidemiology and ecology of simian diseases in relation to man*, Weidenfeld & Nicholsen, London, 86.
127. Winner, H. I. and Hurley, R., *Candida albicans*, Little, Brown, Boston, 1964, 193.
128. Saltarelli, C. G., Gentile, K. A., and Mancuso, S. C., Lethality of *Candida* strains as influenced by the host, *Can. J. Microbiol.*, 21, 648, 1975.
129. Evans, Z. A. and Mardon, D. N., Organ localization in mice challenged with a typical *Candida albicans* strain and a pseudohyphal variant, *Proc. Soc. Exp. Biol. Med.*, 155, 234, 1977.
130. Staib, F., Proteolysis and pathogenicity of *Candida albicans* strains, *Mycopathol. Mycol. Appl.*, 37, 345, 1969.
131. Plempel, M., Antimycotic activity of BAY N 7133 in animal experiments, *J. Antimicrob. Chemother.*, 13, 447, 1984.
132. Bistoni, F., Marconi, P., Frati, L., Bonmassar, E., and Garcia, E., Increase of mouse resistance to *Candida albicans* infection by thymosin, *Infect. Immun.*, 36, 609, 1982.
133. Neta, R. and Salvin, S. B., Mechanisms in the *in vitro* release of lymphokines: relationship of high and low responsiveness to other parameters of the immune response, *Infect. Immun.*, 34, 160, 1981.
134. Rogers, T. J., Balish, E., and Manning, D. D., The role of thymus-dependent cell-mediated immunity in resistance to experimental disseminated candidiasis, *J. Reticuloend.*, 20, 291, 1976.
135. Corbel, M. J. and Eades, S. M., The relative susceptibility of New Zealand Black and CBA mice to infection with opportunic fungal pathogens, *Sabouraudia*, 14, 17, 1976.
136. Morelli, R. and Rosenberg, L. T., Role of complement during experimental *Candida* infection in mice, *Infect. Immun.*, 3, 521, 1971.
137. Hector, R. F., Domer, J. E., and Carrow, E. W., Immune responses to *Candida albicans* in genetically distinct mice, *Infect. Immun.*, 38, 1020, 1982.
138. Cinander, B., Dubiski, S., and Wardlaw, A. C., Distribution, inheritance, and properties of an antigen, MUB1, and its relation to hemolytic complement, *J. Exp. Med.*, 120, 897, 1964.

139. Gelfand, J. A., Hurley, D. L., Fauci, A. S., and Frank, M. M., Role of complement in host defense against experimental disseminated candidiasis, *J. Infect. Dis.*, 138, 9, 1978.

140. Rifkind, D. and Frey, J. A., Influence of gonadectomy on *Candida albicans* urinary tract infection in CFW mice, *Infect. Immun.*, 5, 332, 1972.

141. Schlegel, R. J. and Bellanti, J. A., Increased susceptibility of males to infection, *Lancet*, 2, 826, 1969.

142. Rogers, T. and Balish, E., Experimental *Candida albicans* infection in conventional mice and germ-free rats, *Infect. Immun.*, 14, 33, 1976.

143. Roth, F. J., Jr., Syverton, J. T., and Friedman, J., The effects of roentgen radiation and cortisone upon experimental moniliasis, *Bacteriol. Proc.*, 1952, 87.

144. Gordee, R. S. and Simpson, P. J., Relationships of X irradiation to the enhancement of *Candida albicans* infection, *J. Bacteriol.*, 94, 6, 1967.

145. Louria, D. B., Fallon, N., and Browne, H. G., The influence of cortisone on experimental fungus infections in mice, *J. Clin. Invest.*, 39, 1435, 1960.

146. Meyer, P., Hamberger, H., and Drew, J., Differential effects of ubiquinone Q$_7$ and ubiquinone analogs on macrophage activation and experimental infections in granulocytopenic mice, *Infection*, 8, 256, 1980.

147. Bistoni, F., Baccarini, M., Blasi, E., Marconi, P., Puccetti, P., and Garaci, E., Correlation between in vivo and in vitro studies of modulation of resistance to experimental *Candida albicans* infection by cyclophosphamide in mice, *Infect. Immun.*, 40, 46, 1983.

148. Lopez-Berestein, G., Hopfer, R. L., Mehta, K., Hersh, E. M., and Juliano, R. L., Prophylaxis of *Candida albicans* infection in neutropenic mice with liposome-encapsulated amphotericin B, *Antimicrob. Agents Chemother.*, 25, 366, 1984.

149. Mukherji, A. K. and Basu Mallick, K. C., Disseminated candidiasis in cyclophosphamide induced leucopenic state: an experimental study, *Ind. J. Med. Res.*, 60, 1584, 1972.

150. Linquist, J. A., Rabinovich, S., and Smith, I. M., 5-fluorocytosine in the treatment of experimental candidiasis, *Antimicrob. Agents Chemother.*, 4, 58, 1973.

151. Buhles, W. C., Jr. and Shifrine, M., Adjuvant protection against bacterial infection in granulocytopenic mice, *J. Infect. Dis.*, 136, 90, 1977.

152. Sher, N. A., Chapara, S. D., Greenber, L. E., and Bernard, S., Effects of BCG, *Corynebacterium parvum*, and methanol-extraction residue in the reduction of mortality from *Staphylococcus aureus* and *Candida albicans* infections in immunosuppressed mice, *Infect. Immun.*, 12, 1325, 1975.

153. Moser, S. A. and Domer, J. E., Effects of cyclophosphamide on murine candidiasis, *Infect. Immun.*, 27, 376, 1980.

154. Salvin, S. B. and Tanner, E. P., Resistance and susceptibility to infection in inbred murine strains. III. Effect of thymosin on cellular immune responses of alloxan diabetic mice, *Clin. Exp. Immunol.*, 54, 133, 1983.

155. Robinette, E. H., Jr. and Mardon, D. N., Delayed lethal response to *Candida albicans* infection in mice bearing the Lewis lung carcinoma, *J. Nat. Cancer Inst.*, 55, 731, 1975.

156. Mardon, D. N. and Robinette, E. H., Jr., Organ distribution and viability of *Candida albicans* in noncancerous and tumor-bearing (Lewis lung carcinoma) mice, *Can. J. Microbiol.*, 24, 1515, 1978.

157. Marra, S. and Balish, E., Immunity to *Candida albicans* induced by *Listeria monocytogenes*, *Infect. Immun.*, 10, 72, 1974.

158. Ziegler, A. E., Bicker, U., and Hebold, G., Experimental investigations on increased resistance to infection swith *Candida albicans* and *Staphylococcus aureus* Smith by 4-imino-1,4-diazobicyclo-(3.1.0)-hexane-2-on BM 06.002 (Prop. INN Imexon) in mice, *Exp. Pathol. Bd.*, 14 (Suppl.), 321, 1977.

159. Hurtrel, B., Lagrange, P. H., and Michel, J.-C., Absence of correlation between delayed-type hypersensitivity and protection in experimental systemic candidiasis in immunized mice, *Infect. Immun.*, 31, 95, 1981.

160. Sen, P., Smith, J. K., Buse, M., Hsieh, H. C., Lavenhar, M. A., Linz, D., and Louria, D. B., Modification of an experimental mouse *Candida* infection by human dialyzable leukocyte extract, *Sabouraudia*, 20, 85, 1982.

161. Fraser-Smith, E. B., Epstein, D. A., Larsen, M. A., and Matthews, T. R., Protective effect of a muramyl dipeptide analog encapsulated or mixed with liposomed against *Candida albicans* infection, *Infect. Immun.*, 39, 172, 1983.

162. Stiller, R. L., Bennett, J. E., Scholer, H. J., Wall, M., Polak, A., and Stevens, D. A., Correlation of in vitro susceptibility test results with in vivo response: flucytosine therapy in a systemic candidiasis model, *J. Infect. Dis.*, 147, 1070, 1983.

163. Lopez-Berestein, G., Mehta, R., Hipfer, R. L., Mills, K., Kasi, L., Mehta, K., Fainstein, V., Luna, M., Hersh, E. M., and Juliano, R., Treatment and prophylaxis of disseminated infection due to *Candida albicans* in mice with liposome-encapsulated amphotericin B, *J. Infect. Dis.*, 147, 939, 1983.

164. Tremblay, C., Barza, M., Fiore, C., and Szoka, F., Efficacy of liposome-intercalated amphotericin B in the treatment of systemic candidiasis in mice, *Antimicrob. Agents Chemother.*, 26, 170, 1984.
165. Jevons, S., Gymer, G. E., Brammer, K. W., Cox, D. A., and Leeming, M. R. G., Antifungal activity of tioconazole (UK-20,349), a new imidazole derivative, *Antimicrob. Agents Chemother.*, 15, 597, 1979.
166. Schar, G., Kayser, F. H., and Dupont, M. C., Antimicrobial activity of econazole and miconazole *in vitro* and in experimental candidiasis and aspergillosis, *Chemotherapy*, 22, 211, 1976.
167. Lefler, E. and Stevens, D. A., New azole compounds: vibunazole (Bay n 7133) and Bay l 9136, compared with ketoconazole in the therapy of systemic candidosis and in pharmacokinetic studies, in mice, *J. Antimicrob. Chemother.*, 15, 69, 1985.
168. Chalkley, L. J., Trinci, A. P. J., and Pope, A. M. S., Effect of mycolase and amphotericin B on *Candida albicans* and *Candida pseudotropicalis in vitro* and *in vivo*, *Sabouraudia*, 23, 147, 1985.
169. Polak, A., Scholer, H. J., and Wall, M., Combination therapy of experimental candidiasis, cryptococcosis and aspergillosis in mice, *Chemotherapy*, 28, 461, 1982.
170. Grunberg, E., Titsborth, E., and Bennett, M., Chemotherapeutic activity of 5-fluorocytosine, *Am. Rev. Resp. Dis.*, 84, 504, 1961.
171. Gordee, R. S. and Matthews, T. R., Evaluation of systemic antifungal agents in X-irradiated mice, *Appl. Microbiol.*, 20, 624, 1970.
172. Polak, A., Oxiconazole, a new imidazole derivative. Evaluation of antifungal activity in vitro and in vivo, *Arzneim. Forsch. (Drug Res.)*, 32(I), 17, 1982.
173. Rippon, J. W. and Anderson, D. N., Experimental mycosis in immunosuppressed rabbits. I. Acute and chronic candidiasis, *Mycopathologia*, 64, 91, 1976.
174. MacDonald, F. and Odds, F. C., Inducible proteinase of *Candida albicans* in diagnostic serology and in the pathogenesis of systemic candidiasis, *J. Med. Microbiol.*, 13, 423, 1980.
175. Warren, R. C., Richardson, M. D., and White, L. O., Enzyme-linked immunosorbent assay of antigens from *Candida albicans* circulating in infected mice and rabbits: the role of mannan, *Mycopathologia*, 66, 179, 1978.
176. Eng, R. H., Chmel, H., and Buse, M., Serum levels of arabinitol in the detection of invasive candidiasis in animals and humans, *J. Infect. Dis.*, 143, 677, 1981.
177. Huang, S. Y., Berry, C. W., Newman, J. T., Cooper, W. H., and Zachariah, N. Y., A radioimmunoassay method for the rapid detection of *Candida* antibiodies in experimental systemic candidiasis, *Mycopathologia*, 67, 55, 1979.
178. Repentigny, L., Kuykendall, R. J., Chandler, F. W., Broderson, J. R., and Reiss, E., Comparison of serum mannan, arabinitol, and mannose in experimental disseminated candidiasis, *J. Clin. Microbiol.*, 19, 804, 1984.
179. Harding, S. A., Brody, J. P., and Normansell, D. E., Antigenemia detected by enzyme-linked immunosorbent assay in rabbits with systemic candidiasis, *J. Lab. Clin. Med.*, 95, 959, 1980.
180. Lew, M. A., Siber, C. R., Donahue, D. M., and Maiorca, F., Enhanced detection with an enzyme-linked immunosorbent assay of *Candida* mannan in antibody-containing serum after heat extraction, *J. Infect. Dis.*, 145, 45, 1982.
181. Poor, A. H. and Cutler, J. E., Partially purified antibodies used in a solid-phase radioimmunoassay for detecting candidal antigenemia, *J. Clin. Microbiol.*, 9, 362, 1979.
182. Weiner, M. H. and Coats-Stephen, M., Immunodiagnosis of systemic candidiasis: mannan antigenemia detected by radioimmunoassay in experimental and human infections, *J. Infect. Dis.*, 140, 989, 1979.
183. Wong, B., Bernard, E. M., Gold, J. W. M., Fong, D., Silber, A., and Armstrong, D., Increased arabinitol levels in experimental candidiasis in rats: arabinitol appearance rates, arabinotol/creatinine ratios, and severity of infection, *J. Infect. Dis.*, 146, 346, 1982.
184. Marier, R. L., Milligan, E., and Fan, Y.-D., Elevated mannose levels detected by gas-liquid chromatography in hydrolysates of serum from rats and humans with candidiasis, *J. Clin. Microbiol.*, 16, 123, 1982.
185. Galgiani, J. N. and Van Wyck, D. B., Ornithyl amphotericin methyl ester treatment of experimental candidiasis in rats, *Antimicrob. Agents Chemother.*, 26, 108, 1984.
186. Balk, M. W., Crumrine, M. H., and Fischer, G. W., Evaluation of miconazole therapy in experimental disseminated candidiasis in laboratory rats, *Antimicrob. Agents Chemother.*, 13, 321, 1978.
187. Braude, A. and Rock, J. A., The syndrome of acute disseminated moniliasis in adults, *Arch. Inter. Med.*, 104, 91, 1959.
188. Fuentes, C. A., Schwarz, J., and Aboulafia, R., Some aspects of the pathogenicity of *Candida albicans* in laboratory animals, *Mycopathol. Mycol. Appl.*, 6, 176, 1952.
189. Winner, H. I., Experimental moniliasis in the guinea-pig, *J. Pathol. Bacteriol.*, 79, 420, 1960.
190. Hurley, D. L. and Fauci, A. S., Disseminated candidiasis. I. An experimental model in the guinea pig, *J. Infect. Dis.*, 131, 516, 1975.

191. Fransen, J., Van Cutsem, J., Vandesteene, R., and Janssen, P. A. J., Histopathology of experimental systemic candidosis in guinea-pigs, *Sabouraudia*, 22, 91, 1984.

192. Van Cutsem, J., Fransen, J., and Janssen, P. A. J., Animal models for systemic dermatophyte and *Candida* infection with dissemination to the skin, in *Models in Dermatology, Dermatotoxicology and Dermatopharmacology*, Maibach, H. I. and Lowe, N. J., Eds., S. Karger, New York, 1985.

193. Van Cutsem, J., Fransen, J., Van Gerven, F., and Janssen, P. A. J., Oral treatment with ketoconazole in systemic candidosis of guinea-pigs: microbiology, hematology and histopathology, *Sabouraudia*, 23, 189, 1985.

194. Ruthe, R. C., Andersen, B. R., Cunningham, B. L., and Epstein, R. B., Efficacy of granulocyte transfusions in the control of systemic candidiasis in the leukopenic host, *Blood*, 52, 493, 1978.

195. Epstein, R. B. and Chow, H. S., An analysis of quantitative relationships of granulocyte transfusion therapy in canines, *Transfusion*, 21, 112, 1981.

196. Chow, H. S., Sarpel, S. C., and Epstein, R. B., Experimental candidiasis in neutropenic dogs: tissue burden of infection and granulocyte transfusion effects, *Blood*, 59, 328, 1982.

197. Bayer, A. S., Edwards, J. E., Jr., and Guze, L. B., Experimental intraabdominal candidiasis: macroscopic, microscopic and cultural natural history, *Digestion*, 20, 365, 1980.

198. Demierre, G. and Freedman, L. R., Experimental endocarditis: prophylaxis of *Candida albicans* infections by 5-fluorocytosine in rabbits, *Antimicrob. Agents Chemother.*, 16, 252, 1979.

199. Howard, D. H., Fate of *Histoplasma capsulatum* in guinea pig polymorphonuclear leukocytes, *Infect. Immun.*, 8, 412, 1973.

200. Edwards, J. E., Jr., Montgomerie, J. Z., Foos, R. Y., Shaw, V. K., and Guze, L. B., Experimental hematogenous endophthalmitis caused by *Candida albicans*, *J. Infect. Dis.*, 131, 649, 1975.

201. Tarsi, R., Simonetti, N., and Orpianesi, C., Experimental candidiasis in rabbits: protective action of fructose-1,6-diphosphate, *Mycopathologia*, 81, 111, 1983.

202. Parker, J. C., Jr., Cleary, T. J., and Kogure, K., The effects of transient candidemia on the brain: preliminary observations on a rodent model for experimental deep candidosis, *Surg. Neurol.*, 11, 44, 1979.

203. Parker, J. C., Jr., Cleary, T. J., Monji, T., Kogure, K., and Castro, A., Modifying cerebral candidiasis by altering the infectious entry route, *Arch. Pathol. Lab. Med.*, 104, 537, 1980.

204. Hoffmann, D. H. and Waubke, T. H., Experimental studies of metastatic ophthalmia with *Candida albicans*, *Graefe. Arch. Ophthalmol.*, 164, 174, 1961.

205. Hoffmann, D. H., Die experimentelle endogene Entzundung des Augeninnern durch *Candida albicans*. Ophthalmoskopische, histologische, und mikrobiologische Studien zum Ablau der Infektion beim Kaninchen, *Ophthalmologica*, 151, 1, 1966.

206. Vergara, P., Fuertes, G. A., Buen, S., Ochoa, A. G., and Santos, R., Lesiones oculares en la inoculacion experimental con *Candida albicans* (nota preliminar), *An. Soc. Mexicana Oftalmol. Oto-Rino Laringol.*, 37, 234, 1964.

207. Santos, R., De Buen, S., and Juarez, P., Experimental *Candida albicans* chorioretinitis treated by laser, *Am. J. Ophthalmol.*, 63, 440, 1967.

208. Demant, E. and Easterbrook, M., An experimental model of *Candida* endophthalmitis, *Am. J. Ophthalmol.*, 12, 304, 1977.

209. Cohen, M., Edwards, J. E., Jr., Hensley, and Guze, L. B., Experimental hematogenous *Candida* albicans endophthalmitis: electron microscopy, *Invest. Ophthalmol. Vis. Sci.*, 16, 498, 1977.

210. Henderson, D. K., Edwards, J. E., Jr., Ishida, K., and Guze, L. B., Experimental hematogenous *Candida* endophthalmitis: diagnostic approaches, *Infect. Immun.*, 23, 858, 1979.

211. Edwards, J. E., Jr., Montgomerie, J. Z., Ishida, K., Morrison, H. O., and Guze, L. B., Experimental hematogenous endophthalmitis due to *Candida*: species variation in ocular pathogenicity, *J. Infect. Dis.*, 135, 294, 1977.

212. Fujita, N. K., Henderson, D. K., Hockey, L. J., Guze, L. B., and Edwards, J. E., Jr., Comparative ocular pathogenicity of Cryptococcus neoformans, Candida glabrata, and Aspergillus fumigatus in the rabbits, *Invest. Ophthalmol.*, 22, 410, 1982.

213. Rimbaud, P., Rioux, J. A., and Boulad, L., Une nouvelle localisation des candidoses: la candidose intraoculaire (étude clinique et experimentale), *Ann. Dermatol. Syphiligr. (Paris)*, 90, 135, 1963.

214. Jones, D. B., Chemotherapy of experimental endogenous *Candida albicans* endophthalmitis, *Trans. Am. Ophthal. Soc.*, 78, 846, 1980.

215. Jones, D. B., Green, M. T., Osato, M. S., Broberg, P. H., and Gentry, L. O., Endogenous *Candida albicans* endophthalmitis in the rabbit, *Arch. Ophthalmol.*, 99, 2182, 1981.

216. Axelrod, A. J. and Peyman, G. A., Intravitreal amphotericin B treatment of experiment fungal endophthalmitis, *Am. J. Ophthalmol.*, 76, 584, 1973.

217. Huang, K., Peyman, G. A., and McGetrick, J., Vitrectomy in experimental endophthalmitis. I. Fungal infection, *Ophthal. Surg.*, 10, 84, 1979.

218. McGetrick, J. J., Peyman, G. A., and Nyberg, M. A., Amphotericin B methyl ester: evaluation for intravitreous use in experimental fungal endophthalmitis, *Ophthal. Surg.*, 10, 25, 1979.

219. Richard, M., Peyman, G. A., West, C. S., Hammond, G., and Zweig, K., Toxicity and efficacy of vitrectomy fluids: amphotericin B methyl ester in the treatment of experimental fungal endophthalmitis, *Ophthal. Surg.*, 11, 246, 1980.

220. Smith, E. C., Prickly heat; its aetiology and pathology, *Trans. R. Soc. Trop. Med. Hyg.*, 20, 344, 1927.

221. Karcher, K. H., Experimentelle Untersuchungen zur Pathogenität und Biologischen Wirkung *Candida albicans* an Mensch und Tier, *Arch. Klin. Exp. Dermatol.*, 202, 424, 1956.

222. Delmotte, A., Contribution à l'inoculation expérimentale de *C. albicans, Bull. Group. Intern. Rech. Sci. Stomatol.*, 2, 18, 1959.

223. Maibach, H. I. and Kligman, A. M., The biology of experimental human cutaneous moniliasis (*Candida albicans*), *Arch. Dermatol.*, 85, 233, 1962.

224. Kramer, A., Rödel, R., Krause, R., Weuffen, W., and Kuhne, R., Die experimentelle Hahnenkammkandidose als Modell zur Prüfung von sprosspilzwirksamen Antimykotika, *Ztbl. Gesamte Hyg.*, 20, 509, 1974.

225. Kramer, A., Krause, R., Weuffen, W., and Rödel, B., Erste Erfahrungen mit dem Hahnenkammkandidosetest als in vivo-Prüfmodell für antifungiell wirksame Substanzen, *J. Hyg. Epidemiol. Microbiol. Immunol. (Praha)*, 19, 259, 1975.

226. Schwartzman, R. M., Deubler, M. J., and Dice, P. F., II, Experimentally induced cutaneous moniliasis in the dog, *J. Small Anim. Pract.*, 6, 327, 1965.

227. Maestrone, G. and Semar, R., Establishment and treatment of cutaneous *C. albicans* infection in the rabbit, *Naturwissenshaften*, 2, 1, 1968.

228. Wildfeuer, A., Die experimentelle Bewertung sprosspilzwirksamer Antimykotika unter Einbeziehung der kutanen Candidamykose des Kaninchens als neuem Kriterium, *Mykosen*, 15, 119, 1972.

229. Van Cutsem, J. and Thienpont, D., Experimental cutaneous *Candida albicans* infection in guinea-pigs, *Sabouraudia*, 9, 17, 1971.

230. Sohnle, P. G., Frank, M. M., and Kirkpatrick, C. H., Mechanisms involved in elimination of organisms from experimental cutaneous Candida albicans infection in guinea pigs, *J. Immunol.*, 117, 523, 1976.

231. Ray, T. L. and Wuepper, K. D., Experimental cutaneous candidiasis in rodents, *J. Infect. Dis.*, 66, 29, 1976.

232. Kubo, I., Hori, M., Niitani, H., Yokoyama, I., and Takemoto, T., Studies on the pathogenesis of candidiasis. Studies on candidiasis in Japan, in *Tokyo-Research Committee of Candidiasis*, Education Ministry of Japan, 1961, 145.

233. Giger, D. K., Domer, J. E., and McQuitty, J. T., Jr., Experimental murine candidiasis: pathological and immune responses to cutaneous inoculation with *Candida albicans, Infect. Immun.*, 19, 499, 1978.

234. Ray, T. L. and Wuepper, K. D., Recent advances in cutaneous candidiasis, *Intern. J. Trop. Dermatol.*, 17, 683, 1978.

235. Ray, T. L. and Wuepper, K. D., Experimental cutaneous candidiasis in rodents. II. Role of the stratum corneum barrier and serum complement as a mediator of a protective inflammatory response, *Arch. Dermatol.*, 114, 539, 1978.

236. Domer, J. E. and Moser, S. A., Experimental murine candidiasis: cell-mediated immunity after cutaneous challenge, 20, 88, 1978.

237. Simonetti, N. and Strippoli, V., Pathogenicity of the Y form as compared to M form in experimentally induced *C. albicans* infections, *Mycopathologia*, 51, 19, 1973.

238. Sohnle, P. G. and Kirkpatrick, C. H., Study of possible mechanisms of basophil accumulation in experimental cutaneous candidiasis in guinea pigs, *J. Allerg. Clin. Immunol.*, 59, 171, 1977.

239. Sohnle, P. G. and Kirkpatrick, C. H., Epidermal proliferation in the defense against experimental cutaneous candidiasis, *J. Invest. Dermatol.*, 70, 130, 1978.

240. Cauwenbergh, G. F. M. J., Degreef, H., and Verhowve, S. G. C., Topical ketoconazole in dermatology: a pharmacological and clinical review, *Mykosen*, 27, 395, 1983.

241. Selbie, F. R. and O'Grady, F., A measurable tuberculous lesion in the thigh of the mouse, *Br. J. Exp. Pathol.*, 35, 556, 1954.

242. O'Grady, F., Measurement of drug action in experimental Candida infections, in *Symposium on Candida Infections*, Winner, H. I. and Hurley, R., Eds., E & S. Livingston, London, 1966, 54.

243. Thompson, R. E. M., Effects of antibiotics and steroids on the growth of Candida *in vitro* and *in vivo*, in *Symposium on Candida infections*, Winner, H. I. and Huley, R., Eds., E. & S. Livingston, London, 1966, 65.

244. Pearsall, N. N. and Lagunoff, D., Immunological responses to *Candida albicans*. II. Amyloidosis in mice induced by candidiasis, *Infect. Immun.*, 10, 1397, 1974.

245. Pearsall, N. N., Adams, B. L., and Bunni, R., Immunologic responses to *Candida albicans*. III. Effects of passive transfer of lymphoid cells or serum on murine candidiasis, *J. Infect. Dis.*, 120, 1176, 1978.

246. Johnson, J. A., Lau, B. H. S., Nutter, R. L., Slater, J. M., and Winter, C. E., Effect of L1210 leukemia on the susceptibility of mice to *Candida albicans* infections, *Infect. Immun.*, 19, 146, 1978.

247. Tabeta, H., Mikami, Y., Abe, F., Ommura, Y., and Arai, T., Studies on defense mechanisms against *Candida albicans* infection in congenitally athymic nude (nu/nu) mice, *Mycopathologia*, 84, 107, 1984.

248. Mann, S. J. and Blank, F., Systemic amyloidosis in mice inoculated with lyophilised Candida cells, *Infect. Immun.*, 11, 1371, 1975.

249. Savage, A. and Tribe, C. R., Experimental murine amyloidosis: experience with *Candida albicans* as an amyloidogenic agent and liver biopsy as a diagnostic tool, *J. Pathol.*, 127, 199, 1979.

250. Hollingsworth, J. W. and Carr, J., Experimental candidal arthritis in the rabbit, *Sabouraudia*, 11, 56, 1973.

251. McRipley, R. J., Erhard, P. I., Schwind, R. A., and Whitney, R. R., Evaluation of vaginal antifungal formulation in vivo, *Postgrad. Med. J.*, 55, 648, 1979.

252. Taschdjian, C. L., Reiss, R., and Kozinn, P. J., Experimental vaginal candidiasis in mice; its implications for superficial candidiasis in humans, *J. Invest. Dermatol.*, 34, 89, 1960.

253. Scholer, H. J., Experimentelle vaginal-Candidiasis der Ratte, *Pathol. Microbiol.*, 23, 62, 1960.

254. Sobel, J. D. and Muller, G., Comparison of ketoconazole, Bay N7133, and Bay L9139 in the treatment of experimental vaginal candidiasis, *Antimicrob. Agents Chemother.*, 24, 434, 1983.

255. Thienpont, D., Vam Cutsem, J., Van Neiten J. M., Niemegeers, C. J. E., and Marsboom, R., Biological and toxicological properties of econazole, a broad-spectrum antimycotic, *Arzneim. Forsch. (Drug Res.)*, 25, 224, 1975.

256. Wildfeuer, V. A., Die Chemotherapie der vaginalen Trichomoniasis und Candidosis der Maus, *Arzneim. Forsch. (Drug Res.)*, 24, 937, 1974.

257. Mizuba, S., Muir, R. D., Johnson, F., Lee, K., and Zeicina, S., *In vitro* and *in vivo* studies with methyl partricin, *Dev. Ind. Microbiol.*, 15, 338, 1974.

258. Polak, A., Synergism of polyene antibiotics with 5-fluorocytosine, *Chemotherapy*, 24, 2, 1978.

259. Sobel, J. D., Muller, G., and McCormick, J. F., Experimental chronic vaginal candidosis in rats, *Sabouraudia*, 23, 199—206, 1985.

260. Segal, E., Soroka, A., and Lehrer, N., Attachment of Candida to mammalian tissues — clinical and experimental studies, *Zbl. Bakt. Hyg. A*, 257, 257, 1984.

261. Eng, R. H. K., Chmel, H., and Buse, M., Serum levels of arabinitol in the detection of invasive candidiasis in animals and humans, *J. Infect. Dis.*, 143, 677, 1981.

262. Mackinnon, J. E., Nuevo sentido de variacion en mycotorula (monilia) albicans, *Archos Soc. Biol. Montev.*, 7, 162, 1936.

263. Jones, J. H. and Adams, D., Experimentally induced acute oral candidosis in the rat, *Br. J. Dermatol.*, 83, 670, 1970.

264. Adams, D. and Jones, J. H., Life history of experimentally induced acute oral candidiasis in the rat, *J. Dent. Res.*, 50, 643, 1971.

265. Russell, C. and Jones, J. H., Effects of oral inoculation of *Candida albicans* in tetracycline-treated rats, *J. Med. Microbiol.*, 6, 275, 1973.

266. Jones, J. H. and Russell, C., The histology of chronic candidal infection of the rat's tongue, *J. Pathol.*, 113, 97, 1974.

267. Jones, J. H. and Russell, C., The effects of inoculation of the yeast and mycelial phases of *Candida albicans* in rats fed normal and carbohydrate rich diets, *Arch. Oral Biol.*, 18, 409, 1973.

268. Jones, J. H., Russell, C., Young, C., and Owen, D., Tetracycline and the colonization and infection of the mouths of germ-free and conventionalized rats with *Candida albicans*, *J. Antimicrob. Chemother.*, 2, 247, 1976.

269. Russell, C., Jones, J. H., and Gibbs, A. C. C., The carriage of *Candida albicans* in the mouths of rats treated with tetracycline briefly or for a prolonged period, *Mycopathologia*, 58, 125, 1976.

270. Fisker, A. V., Schiott, S. R., and Philipsen, H. P., Short-term oral candidosis in rats, with special reference to the site of infection, *Acta Pathol. Microbiol. Scand. Sect. B*, 90, 49, 1982.

271. Allen, C. M. and Beck, F. M., Strain-related differences in pathogenicity of *Candida albicans* for oral mucosa, *J. Infect. Dis.*, 147, 1936, 1983.

272. Balish, E. and Phillips, A. W., Growth, morphogenesis and virulence of *Candida albicans* after oral inoculation in the germ-free and conventional chick, *J. Bacteriol.*, 91, 1736, 1966.

273. Phillips, A. W. and Balish, E., Growth and invasiveness of *Candida albicans* in the germ-free and conventional mouse after oral challenge, *Appl. Microbiol.*, 14, 737, 1966.

274. Winner, H. I. and Hurley, R., *Candida Albicans*, Churchill, London, 1964.

275. Clark, J. D., Influence of antibiotics or certain intestinal bacteria on orally administered *Candida albicans* in germ-free and conventional mice, *Infect. Immun.*, 4, 731, 1971.

276. Nishikawa, T., Hatano, H., Ohnishi, N., Sakaki, S., and Nonomura, T., Establishment of *Candida albicans* in the alimentary tract of germ-free mice and antagonism with *Escherichia coli* after oral inoculation, *Jpn. J. Microbiol.*, 13, 263, 1969.

277. Turner, J. R., Butler, T. F., Johnson, M. E., and Gordee, R. S., Colonization of the intestinal tract of conventional mice with *Candida albicans* and treatment with antifungal agents, *Antimicrob. Agents Chemother.*, 9, 787, 1976.

278. Helstrom, P. B. and Balish, E., Effect of oral tetracycline, the microbial flora, and the athymic state on gastrointestinal colonization and infection of BALB/c mice with *Candida albicans*, *Infect. Immun.*, 23, 764, 1979.

279. Auger, P. and Joly, J., Étude de quelques facteurs intervenant dans la colonisation du tube digestif de la souris blanche par le Candida albicans, *Can. J. Microbiol.*, 22, 334, 1976.

280. Wildfeuer, A., Experimentelle Sprosspilzinfektion des Gastrointestinaltraktes der Maus, *Mykosen*, 21, 157, 1978.

281. Myerowitz, R. L., Gastrointestinal and disseminated candidiasis, *Arch. Pathol. Lab. Med.*, 105, 138, 1981.

282. Umenai, T., Konno, S., Yamaguchi, A., Iimura, Y., and Fujimoto, H., Growth of *Candida* in the upper intestinal tract as a possible source of systemic candidiasis, *Tohoku J. Exp. Med.*, 130, 101, 1980.

283. Wingard, J. R., Dick, J. D., Merz, W. G., Sandford, G. R., Saral, R., and Burns, W. H., Pathogenicity of *Candida tropicalis* and *Candida albicans* after gastrointestinal inoculation in mice, *Infect. Immun.*, 29, 808, 1980.

284. Pope, L. M., Cole, G. T., Guentzel, M. N., and Berry, L. J., Systemic and gastrointestinal candidiasis of infant mice after intragastric challenge, *Infect. Immun.*, 25, 702, 1979.

285. Field, L. H., Pope, L. M., Cole, G. T., Guentzel, M. N., and Berry, L. J., Persistence and spread of *Candida albicans* after intragastric inoculation of infant mice, *Infect. Immun.*, 31, 783, 1981.

286. DeMaria, A., Buckley, H., and Von Lichtenberg, F., Gastrointestinal candidiasis in rats treated with antibiotics, cortisone, and azathioprine, *Infect. Immun.*, 13, 1761, 1976.

287. Burke, V. and Gracey, M., An experimental model of gastrointestinal candidiasis, *J. Med. Microbiol.*, 13, 103, 1980.

288. Kurotchkin, T. J. and Lim, C. E., Experimental bronchomoniliasis in sensitized rabbits, *Proc. Soc. Exp. Biol. Med.*, 21, 332, 1963.

289. Evans, W. E. D. and Winner, H. I., The histogenesis of the lesions in experimental moniliasis in rabbits, *J. Pathol. Bacteriol.*, 67, 531, 1954.

290. Felisati, D., Bastianini, L., and De Mitri, T., Antibiotics and *Candida albicans* given by endobranchial route, *Antibiot. Chemother.*, 9, 744, 1959.

291. Damodaran, V. N. and Chakravarty, S. C., Mechanism of production of Candida lesions in rabbits, *J. Med. Microbiol.*, 6, 287, 1973.

292. Urso, B. and Capocaccia, L., Observazioni su un Caso di moniliasi bronchiale can guarde asmatico, *Arch. Ital. Sci. Med. Trop. Parassit.*, 31, 77, 1952.

293. Vogel, R. A. and Krehl, W., Experimental sensitization of guinea pigs with *Candida albicans* and adjuvants, *Am. Res. Tuberc.*, 76, 692, 1957.

294. Jaiswal, A. K., Biochemical studies on pulmonary candidiasis in guinea pigs, *Toxicon*, 19, 570, 1981.

295. Jaiswal, A. K., Phospholipid changes in guinea pig lungs in pulmonary candidiasis, *Toxicon*, 19, 910, 1981.

296. Zaidi, S. H., Shanker, R., and Dogra, R. K. S., Experimental infective pneumoconiosis: effect of asbestos dust and *Candida albicans* infection on the lungs of rhesus monkeys, *Environ. Res.*, 6, 274, 1973.

297. Nugent, K. M. and Onofrio, J. M., Pulmonary tissue resistance to *Candida albicans* in normal and in immunosuppressed mice, *Am. Rev. Resp. Dis.*, 128, 909, 1983.

298. Aoyama, S. and Ata, S., Studies on candidiasis in Japan, Education Ministry of Japan, 1961, 115.

299. Garrison, P. K. and Freedman, L. R., Experimental endocarditis. I. Staphylococcal endocarditis in rabbits resulting from placement of a polyethylene catheter in the right side of the heart, *Yale J. Biol. Med.*, 42, 394, 1970.

300. Freedman, L. R. and Johnson, M. L., Experimental endocarditis. IV. Tricuspid and aortic valve infection with *Candida albicans* in rabbits, *Yale J. Biol. Med.*, 45, 163, 1972.

301. Demierre, G. and Freedman, L. R., Experimental endocarditis: prophylaxis of *Candida albicans* infections by 5-fluorocytosine in rabbits, *Antimicrob. Agents Chemother.*, 16, 252, 1979.

302. Sande, M. A., Bowman, C. R., and Calderone, R. A., Experimental *Candida albicans* endocarditis: characterization of the disease and response to therapy, *Infect. Immun.*, 17, 140, 1977.

303. Calderone, R. A., Rotondo, M. F., and Sande, M. A., *Candida albicans* endocarditis: ultrastructural studies of vegetation formation, *Infect. Immun.*, 20, 279, 1978.

304. Sande, M. A., Evaluation of antimicrobial agents in the rabbit model of endocarditis, *Rev. Infect. Dis.*, 3, S240, 1981.

305. Kwon-Chung, K. J., A new genus *Filobasidiella*, the perfect state of *Cryptococcus neoformans*, *Mycologia*, 67, 1197, 1975.

306. Kwon-Chung, K. J., A new species of *Filobasidiella*, the sexual state of *Cryptococcus neoformans* B and D serotypes, *Mycologia*, 68, 942, 1976.

307. Emmons, C. W., Isolation of *Cryptococcus neoformans* from soil, *J. Bacteriol.*, 62, 685, 1954.

308. Staib, F., *Cryptococcus neoformans* bein Kanarienvogel, *Zentral. Bakteriol. Parasitenkd. Infektionskn. Hyg., Abt.*, 1, *Orig., Reihe A*, 185, 129, 1962.

309. Campbell, G. D., Primary pulmonary cryptococcosis, *Am. Rev. Resp. Dis.*, 94, 236, 1966.

310. Gordonson, J., Birnbaum, W., Jacobson, G., and Sargent, E. N., Pulmonary cryptococcosis, *Radiology*, 112, 557, 1974.

311. Riley, D. J. and Edelman, N. H., Hemoptysis in pulmonary cryptococcosis, *J. Med. Soc. N.J.*, 75, 553, 1978.

312. Duperval, R., Hermans, P. E., Brewer, N. S., and Roberts, G. D., Cryptococcosis, with emphasis on the significance of isolation of *Cryptococcus neoformans* from the respiratory tract, *Chest*, 72, 13, 1977.

313. Farhi, F., Bulmer, G. S., and Tacker, J. R., *Cryptococcus neoformans*. IV. The not-so-encapsulated yeast, *Infect. Immun.*, 1, 526, 1970.

314. Hatcher, C. R., Jr., Sehdeva, J., Waters, W. C., III, Schulze, Z., Logan, W. D., Jr., Symbas, P., and Abbott, O. A., Primary pulmonary cryptococcosis, *J. Thorac. Cardiovasc. Surg.*, 61, 39, 1971.

315. Katz, R. I., Birnbaum, H., and Eckman, B. H., Resection of pulmonary cryptococcosis associated with meningitis, *Am. Rev. Resp. Dis.*, 84, 725, 1961.

316. Littman, M. L. and Walters, J. E., Cryptococcosis: current status, *Am. J. Med.*, 45, 922, 1968.

317. Haugen, R. K. and Baker, R. D., The pulmonary lesions in cryptococcosis with special reference to subpleural nodes, *Am. J. Clin. Pathol.*, 24, 1381, 1954.

318. Emmons, C. W., Binford, C. H., Utz, J. P., and Dwon-Chung, K. J., *Medical Mycology*, 3rd ed., Lea & Febiger, Philadelphia, 1977, 206.

319. Lewis, J. L. and Rabinovich, S., The wide spectrum of cryptococcal infections, *Am. J. Med.*, 53, 315, 1972.

320. Baker, R. D., The pathologic anatomy of mycoses, in *Handbuch der speziellen pathologischen Anatomie und Histologie*, Vol. 5 (Part 3), Uelinger, E., Ed., Springer-Verlag, Heidelberg, 1971.

321. Hays, R. J., Mackensie, D. W. R., Campbell, C. K., and Philpot, C. M., Cryptococcosis in the United Kingdom and the Irish Republic: an analysis of 69 cases, *J. Infect.*, 2, 13, 1980.

322. Diamond, R. D. and Bennett, J. E., Prognostic factors in cryptococcal meningitis, *Ann. Int. Med.*, 80, 176, 1974.

323. Barron, C. N., Cryptococcosis in animals, *J. Am. Vet. Med. Assoc.*, 127, 125, 1955.

324. Graybill, J. R. and Taylor, R. L., Host defense in cryptococcosis. I. An in vivo model for evaluating host defense, *Int. Arch. Allergy Appl. Immunol.*, 57, 101, 1978.

325. Bennett, J. E., Cryptococcosis, in *Infectious Diseases*, Hoeprich, P. D., Ed., Harper & Row, New York, 1974, 945.

326. Bennett, J. E. and Hasenclever, H. F., *Cryptococcus neoformans* polysaccharide: studies of serologic properties and role in infection, *J. Immunol.*, 94, 916, 1965.

327. Bodenhoff, J., Chronic cryptococcosis in the mouse, *Acta Pathol. Microbiol. Scand.*, 75, 169, 1969.

328. Cauley, L. K. and Murphy, J. W., Response of congenitally athymic (nude) and phenotypically normal mice to a Cryptococcus neoformans infection, *Infect. Immun.*, 23, 644, 1979.

329. Dykstra, M. A. and Friedman, L., Pathogenesis, lethality, and immunizing effect of experimental cutaneous cryptococcosis, *Infect. Immun.*, 2, 446, 1978.

330. Cordon, M. A. and Vedder, D. K., Serological tests in diagnosis and prognosis of cryptococcosis, *J. Am. Med. Assoc.*, 174, 131, 1966.

331. Hay, R. J. and Reiss, E., Delayed-type hypersensitivity responses in infected mice elicited by cytoplasmic fractions of *Cryptococcus neoformans*, *Infect. Immun.*, 22, 72, 1978.

332. Karaoui, R. M., Hall, N. K., and Larsh, H. W., Role of macrophages in immunity and pathogenesis of experimental cryptococcosis induced by the airborne route. I. Pathogenesis and acquired immunity of *Cryptococcus neoformans*, *Mykosen*, 20, 380, 1977.

333. Perceval, A. K., Experimental cryptococcosis: hypersensitivity and immunity, *J. Pathol. Bacteriol.*, 89, 645, 1965.

334. Ritter, R. C. and Larsh, H. W., The infection of white mice following an intranasal instillation of *Cryptococcus neoformans*, *Am. J. Hyg.*, 78, 241, 1963.

335. Smith, C. D., Ritter, R., Larsh, H. W., and Furcolow, M. L., Infection of white Swiss mice with airborne *Cryptococcus neoformans, J. Bacteriol.,* 87, 1364, 1964.

336. Kong, Y. M. and Levine, H. B., Experimentally induced immunity in the mycoses, *Bacteriol. Rev.,* 31, 35, 1967.

337. Pappagianis, D., Miller, R. L., Smith, C. E., and Kobayashi, G. S., Immunization of mice with viable *Coccidioides immitis, Am. Rev. Resp. Dis.,* 82, 244, 1960.

338. Rhodes, J. C., Wicker, L. S., and Urba, W. J., Genetic control of susceptibility to *Cryptococcus neoformans* in mice, *Infect. Immun.,* 29, 494, 1980.

339. Duke, S. S. and Fromtling, R. A., Effects of diethylstilbestrol and cyclophosphamide on the pathogenesis of experimental *Cryptococcus neoformans* infections, *Sabouraudia,* 22, 125, 1984.

340. Monga, D. P., Kumar, R., Mohapatra, L. N., and Malaviya, A. N., Experimental cryptococcosis in normal and B-cell-deficient mice, *Infect. Immun.,* 26, 1, 1979.

341. Rhodes, J. C., Contribution of complement component C5 to the pathogenesis of experimental murine cryptococcosis, *Sabouraudia,* 23, 225, 1985.

342. Sneller, M. R., Hariri, A., Sorenson, W. G., and Larsh, H. W., Comparative study of trichothecin, amphotericin B, and 5-fluorocytosine against *Cryptococcus neoformans in vitro* and *in vivo, Antimicrob. Agents Chemother.,* 12, 390, 1977.

343. Levine, S., Zimmerman, H. M., and Scorza, A., Experimental cryptococcosis (torulosis), *Am. J. Pathol.,* 33, 385, 1957.

344. Fazekas, C. and Schwarz, J., Histology of experimental murine cryptococcosis, *Am. J. Pathol.,* 34, 517, 1958.

345. Bergman, F., Pathology of experimental cryptococcosis. A study of course and tissue response in subcutaneously induced infection in mice, *Acta Pathol. Microbiol. Scand.,* 51(Suppl. 147), 1, 1961.

346. Grosse, G., Mishra, S. K., and Staib, F., Selective involvement of the brain in experimental murine cryptococcosis. II. Histopathological observations, *Zentral. Bakteriol. Parasitenkd. Infektionskn. Hyg., Abt. 1: Orig., Reihe A,* 233, 106, 1975.

347. Nishimura, K. and Miyaji, M., Histopathological studies on experimental cryptococcosis in nude mice, *Mycopathologia,* 68, 145, 1979.

348. Miyaji, M. and Nishimura, K., Experimental cryptococcosis in nude mice, *Excerpta Medica, I.C.S.,* Amsterdam, 1980, 75.

349. Miyaji, M. and Nishimura, K., Studies on organ specificity in experimental murine cryptococcosis, *Mycopathologia,* 76, 145, 1981.

350. Watabe, T., Miyaji, M., and Nishimura, K., Studies on relationship between cysts and granulomas in murine cryptococcosis, *Mycopathologia,* 86, 113, 1984.

351. Bulmer, G. S. and Sans, M. D., *Cryptococcus neoformans* II. Phagocytosis by human leukocytes, *J. Bacteriol.,* 94, 1480, 1967.

352. Mitchell, T. G. and Friedman, L., *In vitro* phagocytosis and intracellular fate of variously encapsulated strains of *Cryptococcus neoformans, Infect. Immun.,* 5, 491, 1972.

353. Neilson, J. B., Ivey, M. H., and Bulmer, G. S., *Cryptococcus neoformans* pseudohyphal forms surviving culture with *Acanthamoeba polyphaga, Infect. Immun.,* 20, 262, 1978.

354. Staib, F. and Mishra, S. K., Selective involvement of the brain in experimental murine cryptococcosis. I. Microbiological observations, *Zentralbl. Parasitenkd. Infektionskr. Bakteriol. Hyg. Abt. 1, Orig. Reihe A,* 232, 355, 1975.

355. Fromtling, R. A., Blackstock, R., and Bulmer, S., Immunization and passive transfer of immunity in murine cryptococcosis, *Excerpta Medica, I.C.S.,* Amsterdam, 1980, 122.

356. Fromtling, R. A., Fromtling, A. M., Staib, F., and Muller, S., Effect of uremia on lymphocyte transformation and chemiluminescence by spleen cells of normal and *Cryptococcus neoformans*-infected mice, *Infect. Immun.,* 32, 1073, 1981.

357. Staib, F., Mishra, S. K., Grosse, G., and Abel, Th., Ocular cryptococcosis — experimental and clinical observations, *Zentralbl. Parasitenkd. Infektionskr. Bakteriol. Hyg. Abt. 1, Orig. Reihe A,* 237, 378, 1977.

358. Reiss, F. and Alture-Werber, E., Immunization of mice with a mutant of *Cryptococcus neoformans, Dermatologica,* 152, 16, 1976.

359. Gordee, R. S. and Matthews, T. R., Evaluation of systemic antifungal agents in X-irradiated mice, *Appl. Microbiol.,* 20, 624, 1970.

360. Fromentin, H., Slazar-Mejicanos, S., and Mariat, F., Experimental cryptococcosis in mice treated with diacetoxyscirpenol, a mycotoxin of *Fusarium, Sabouraudia,* 19, 311, 1981.

361. Adamson, D. M. and Cozad, G. C., Effect of antilymphocyte serum on animals experimentally infected with Histoplasma capsulatum or *Cryptococcus neoformans, J. Bacteriol.,* 100, 1271, 1969.

362. Graybill, J. R. and Mitchell, L., Cyclophosphamide effects on murine cryptococcosis, *Infect. Immun.,* 21, 674, 1978.

363. Craven, P. C., Graybill, J. R., and Jorgenson, J. H., Ketoconazole therapy of murine cryptococcal meningitis, *Am. Rev. Resp. Dis.*, 125, 696, 1982.

364. Graybill, J. R., Kaster, S. R., and Drutz, D. J., Comparative activities of Bay n7133, ICI 153,066, and ketoconazole in murine cryptococcosis, *Antimicrob. Agents Chemother.*, 24, 229, 1983.

365. Craven, P. C. and Graybill, J. R., Combination of oral flucytosine and ketoconazole as therapy for experimental cryptococcal meningitis, *J. Infect. Dis.*, 149, 584, 1984.

366. Graybill, J. R. and Ahrens, J., R 51211 (itraconazole) therapy of murine cryptococcosis, *Sabouraudia*, 22, 445, 1984.

367. Reed, L. J. and Muench, H., A simple method for estimating 50% end point, *Am. J. Hyg.*, 27, 493, 1938.

368. Graybill, J. R., Mitchell, L., and Levine, H. B., Treatment of experimental murine cryptococcosis: a comparison of miconazole and amphotericin B, *Antimicrob. Agents Chemother.*, 13, 277, 1978.

369. Lim, T. S., Murphy, J. W., and Cauley, L. K., Host-etiological agent interactions in intranasally and intraperitoneally induced cryptococcosis in mice, *Infect. Immun.*, 29, 633, 1980.

370. Iwen, P. C., Miller, N. G., and McFadden, H. W., Jr., Treatment of murine pulmonary cryptococcosis with ketoconazole and amphotericin B, *J. Infect. Dis.*, 149, 650, 1984.

371. Henderson, D. W., An apparatus for the study of airborne infection, *J. Hyg.*, 50, 53, 1952.

372. Wilson, J. W., *Clinical and Immunologic Aspects of Fungus Diseases*, Charles C Thomas, Springfield, 1957, 170.

373. Abrahams, I. and Gilleran, T. G., Studies on actively acquired resistance to experimental cryptococcosis in mice, *J. Immunol.*, 85, 629, 1960.

374. Louria, D. B., Kaminski, T., and Findel, G., Further studies on immunity in experimental cryptococcosis, *J. Exp. Med.*, 117, 509, 1961.

375. Louria, D. B., Specific and non-specificity immunity in experimental cryptococcosis in mice, *Am. Rev. Resp. Dis.*, 111, 643, 1960.

376. Williams, M. M., Krick, J. A., and Remington, J. S., Pulmonary infection in the compromised host. I, *Am. Rev. Resp. Dis.*, 114, 359, 1974.

377. Fromtling, R. A., Blackstock, R., Hall, N. K., and Bulmer, G. S., Immunization of mice with an avirulent pseudohyphal form of *Cryptococcus neoformans*, *Mycopathologia*, 68, 179, 1979.

378. Moser, S. A., Lyon, F. L., Domer, J. E., and Williams, J. E., Immunization of mice by intracutaneous inoculation with viable virulent *Cryptococcus neoformans*: immunological and histopathological parameters, *Infect. Immun.*, 35, 685, 1982.

379. Gadebusch, H. H., Passive immunization against *Cryptococcus neoformans*, *Proc. Soc. Exp. Biol. Med.*, 98, 611, 1958.

380. Gadebusch, H. H. and Gikas, P. W., Natural host resistance to infection with *Cryptococcus neoformans.* II. The influence of thiamine on experimental infection in mice, *J. Infect. Dis.*, 112, 125, 1963.

381. Gordon, M. A. and Lapa, E., Anticryptococcal effect in normal and immune serum, in Annual Report of Laboratory Research, New York State Department of Health, Albany, 1964, 71.

382. Louria, D. B. and Kkaminski, T., Passively acquired immunity in experimental cryptococcosis, *Sabouraudia*, 4, 80, 1965.

383. Gordon, M. B., Experimental murine cryptococcosis: effects of hyperimmunization to capsular polysaccharide, *J. Immunol.*, 98, 914, 1967.

384. Effects of stimulation and suppression of cell-mediated immunity on experimental cryptococcosis, *Infect. Immun.*, 17, 187, 1977.

385. Graybill, J. R. and Drutz, D. J., Host defense in cryptococcosis. II. Cryptococcosis in nude mouse, *Cell. Immunol.*, 40, 263, 1978.

386. Graybill, J. R., Mitchell, L., and Drutz, D. J., Host defense on cryptococcosis. III. Protection of nude mice by thymus transplantation, *J. Infect. Dis.*, 140, 546, 1979.

387. Monga, D. P., Kumar, R., Mohapatra, L. N., and Malaviya, A. N., Experimental cryptococcosis in normal and T cell deficient mice, *Ind. J. Med. Res.*, 72, 641, 1980.

388. Murphy, J. W., Gregory, J. A., and Larsh, H. W., Skin testing of guinea pigs and footpad testing of mice with a new antigen for detecting delayed hypersensitivity to Cryptococcus neoformans, *Infect. Immun.*, 9, 404, 1974.

389. Abrahams, I., Further studies on acquired resistance to murine cryptococcosis: enhancing effect of Bordetella pertussis, *J. Immunol.*, 96, 525, 1966.

390. Gentry, L. O. and Remington, J. S., Resistance against *Cryptococcus neoformans* conferred by intracellular bacteria and protozoa, *J. Infect. Dis.*, 123, 22, 1971.

391. Monga, D. P., Role of macrophages in resistance of mice to experimental cryptococcosis, *Infect. Immun.*, 32, 975, 1981.

392. Graybill, J. R. and Mitchell, L., Host defense in cryptococcosis. III. In vivo alteration of immunity, *Mycopathologia*, 69, 171, 1979.

393. Lim, T. S. and Murphy, J. W., Transfer of immunity to cryptococcosis by T-enriched splenic lymphocytes from *Cryptococcus neoformans*-sensitized mice, *Infect. Immun.*, 30, 5, 1980.
394. Diamond, R. D., May, J. E., Kane, M., Frank, M. M., and Bennett, J. E., The role of late complement components and the alternate complement pathway in experimental cryptococcosis, *Proc. Soc. Exp. Biol. Med.*, 144, 312, 1973.
395. Diamond, R. D., May, J. E., Kane, M. A., Frank, M. M., and Bennett, J. E., The role of the classical and alternate complement pathways in host defenses against *Cryptococcus neoformans* infection, *J. Immunol.*, 112, 2260, 1974.
396. Gadebusch, H. H., Natural host resistance to infection with *Cryptococcus neoformans.* I. The effect of the properdin system on the experimental disease, *J. Infect. Dis.*, 109, 147, 1961.
397. Graybill, J. R. and Ahrens, J., Immunization and complement interaction in host defense against murine cryptococcosis, *J. Reticuloendothel Soc.*, 30, 347, 1981.
398. Kozel, T. R., Highison, B., and Stratton, C. J., Localization on encapsulated *Cryptococcus neoformans* of serum components opsonic for phagocytosis by macrophages and neutrophils, *Infect. Immun.*, 43, 574, 1984.
399. Laxalt, K. A. and Kozel, T. R., Chemotaxigenesis and activation of the alternative complement pathway by encapsulated and non-encapsulated Cryptococcus neoformans, *Infect. Immun.*, 26, 435, 1979.
400. Macher, A. M., Bennett, J. E., Gadek, J. E., and Frank, M. M., Complement depletion in cryptococcal sepsis, *J. Immunol.*, 120, 1686, 1978.
401. Fromtling, R. A. and Shadomy, H. J., Immunity in cryptococcosis: an overview, *Mycopathologia*, 77, 183, 1982.
402. Griffin, F. M., Roles of macrophage Fc and C3b receptors in phagocytosis of immunologically coated *Cryptococcus neoformans, Proc. Nat. Acad. Sci. U.S.A.*, 78, 3853, 1981.
403. Grungerg, E., Titsworth, E., and Bennett, M., Chemotherapeutic activity of 5-fluorocytosine, *Antimicrob. Agents Chemother.*, 1963, 566, 1964.
404. Polak, A., Oxiconazole, a new imidazole derivative, *Arzneim. Forsch. (Drug Res.)*, 32(I), 17, 1982.
405. Hamilton, J. D. and Elliott, D. M., Combined activity of amphotericin B and 5-fluorocytosine against *Cryptococcus neoformans* in vitro and in vivo in mice, *J. Infect. Dis.*, 131, 129, 1975.
406. Polak, A., Scholer, H. J., and Wall, M., Combination therapy of experimental candidiasis, cryptococcosis and aspergillosis in mice, *Chemotherapy*, 28, 461, 1982.
407. Plempel, M., Antimycotic activity of BAY N 7133 in animal experiments, *J. Antimicrob. Chemother.*, 13, 447, 1984.
408. Graybill, J. R., Craven, P. C., Mitchell, L. F., and Drutz, D. J., Interaction of chemotherapy and immune defenses in experimental murine cryptococcosis, *Antimicrob. Agents Chemother.*, 14, 659, 1978.
409. Graybill, J. R., Williams, D. M., Van Cutsem, E., and Drutz, D. J., Combination therapy of experimental histoplasmosis and cryptococcosis with amphotericin B and ketoconazole, *Rev. Infect. Dis.*, 2, 551, 1980.
410. Littman, M. L. and Zimmerman, L. E., *Cryptococcosis,* Grune & Stratton, New York, 1956, 71.
411. Strippoli, V., Simonetti, N., and Cassone, A., Effect of a tetracycline antibiotic on the experimental pathogenicity of *Cryptococcus neoformans, Chemotherapy*, 24, 290, 1978.
412. Weiss, C., Perry, I. H., and Shevky, M. C., Infection of the human eye with *Cryptococcus neoformans (Torula histolytica; Cryptococcus hominis):* a clinical and experimental study with a new diagnostic method, *Arch. Ophthalmol.*, 39, 739, 1948.
413. Felton, F. G., Maldonado, W. E., Muchmore, H. G., and Rhoades, E. R., Experimental cryptococcal infection in rabbits, *Am. Rev. Resp. Dis.*, 94, 589, 1966.
414. Perfect, J. R., Lagn, S. D. R., and Durack, D. T., Chronic cryptococcal meningitis. A new experimental model in rabbits, *Am. J. Pathol.*, 101, 177, 1980.
415. Bloomfield, N., Gordon, M. A. and Elmendorf, D. F., Detection of *Cryptococcus neoformans* antigen in body fluids by latex particle agglutination, *Proc. Soc. Exp. Biol. Med.*, 114, 64, 1963.
416. Kahn, M. J., Myer, R., and Koshy, G., Pulmonary cryptococcosis, *Dis. Chest.*, 36, 656, 1959.
417. Levine, S., Zimmerman, H. M., and Scorza, A., Experimental cryptococcosis (torulosis), *Am. J. Pathol.*, 33, 385, 1957.
418. Sethi, K. K., Salfelder, K., and Schwarz, J., Experimental cutaneous primary infection with *Cryptococcus neoformans* (Sanfelice) Vuillemin, *Mycopathol. Mycol. Appl.*, 27, 357, 1965.
419. Wade, L. J. and Stevenson, L. D., *Torula* infection, *Yale J. Med. Biol.*, 13, 467, 1941.
420. Song, M. M., Experimental cryptococcosis of the skin, *Sabouraudia*, 12, 133, 1971.
421. Domer, J. E., Lyon, F. L., and Murphy, J. W., Cellular immunity in a cutaneous model of cryptococcosis, *Infect. Immun.*, 40, 1052, 1983.
422. Perfect, J. R., Lang, S. D. R., and Durack, D. T., Influence of agglutinating antibody in experimental cryptococcal meningitis, *Br. J. Exp. Pathol.*, 62, 595, 1981.

423. Perfect, J. R. and Durack, D. T., Treatment of experimental cryptococcal meningitis with amphotericin B, 5-fluorocytosine, and ketoconazole, *J. Infect. Dis.*, 146, 429, 1982.

424. Graybill, J. R., Ahrens, J., Nealon, T., and Paque, R., Pulmonary cryptococcosis in the rat, *Am. Rev. Resp. Dis.*, 127, 636, 1983.

425. Gadebusch, H. H. and Gikas, P. W., The effect of cortisone upon experimental pulmonary cryptococcosis, *Am. Rev. Resp. Dis.*, 92, 64, 1965.

426. Fujita, N. K., Henderson, D. K., Hockey, L. J., Guze, L. B., and Edwards, J. E., Jr., Comparative ocular pathogenicity of *Cryptococcus neoformans, Candida glabrata*, and *Aspergillus fumigatus* in the rabbit, *Invest. Ophthalmol. Vis. Sci.*, 22, 410, 1982.

427. Kligman, A. M. and Weidman, F. D., Experimental studies on the treatment of human torulosis, *Arch. Derm. Syph.*, 60, 726, 1949.

428. Kreibig, W., Beiderseitige metastatische Ophthalmie durch Blastomyceten, *Klin. Mbl. Augenheilk.*, 104, 64, 1940.

429. Kao, C. J. and Schwarz, J., The isolation of *Cryptococcus* neoformans from pigeon nests, *Am. J. Clin. Pathol.*, 27, 652, 1957.

430. Littman, M. L. and Zimmerman, L. E., Cryptococcosis, Grune & Stratton, New York, 1956, 205.

431. Vanbreuseghem, R., White mice sensitivity to *Cryptococcus neoformans.* Production of resistance against a cerebral inoculation, *Ann. Doc. Belge. Med. Trop.*, 47, 281, 1967.

432. Blouin, P. and Cell, R. M., Experimental ocular cryptococcosis. Preliminary studies in cats and mice, *Invest. Ophthalmol. Vis. Sci.*, 19, 21, 1980.

433. Nagai, T., Kawai, C., and Kumagai, N., Experimental cryptococcal myocarditis, *Jpn. Circ. J.*, 43, 450, 1979.

434. Green, J. R. and Bulmer, G. S., Gastrointestinal inoculation of *Cryptococcus neoformans* in mice, *Sabouraudia*, 17, 233, 1979.

435. Thygeson, P. and Okumoto, M., Keratomycosis: a preventable disease, *Trans. Am. Acad. Ophthalmol. Otolaryngol.*, 78, 433, 1974.

436. Zimmerman, L. E., Mycotic keratitis, *Lab. Invest.*, 11, 1151, 1962.

437. Jones, B. R., Principles in the management of oculomycosis, *Am. J. Ophthal.*, 79, 719, 1975.

438. Ishibashi, Y., Keratomycosis in Japan reported from 1976 to 1980, *Acta Soc. Ophthalmol. Japon.*, 86, 651, 1982.

439. Emmons, C. W., Binford, C. H., Utz, J. P., and Kwong-Chung, K. J., *Medical Mycology*, 3rd ed., Lea & Febiger, Philadelphia, 1977.

440. Romano, A. E., Segal, E., Stein, R., and Eylen, E., Yeasts in banal external ocular inflammation, *Ophthalmologica*, 170, 13, 1975.

441. Tandon, R. N., Wahab, S., and Srivastave, O. P., Experimental infection by Candida krusei (Cast.) Berkhout isolated from a case of corneal ulcer and its sensitivity to antimycotics, *Mykosen*, 27, 355, 1984.

442. François, J. and Rysselaere, M., *Oculomycosis*, Thomas, Springfield, 1972.

443. Chandler, F. W., Kaplan, W., and Ajello, L., Mycotic keratitis, in *A Colour Atlas and Textbook of the Histopathology of Mycotic Diseases*, Wolfe Medical, London, 1980, 83.

444. Nytyananda, K., Sirasubramaniam, P., and Ajello, L., Mycotic keratitis caused by *Curvularia lunata, Sabouraudia*, 2, 35, 1962.

445. Wind, C. A. and Plack, F. M., Keratomycosis due to *Curvularia lunata, Arch. Ophthalmol.*, 84, 694, 1970.

446. Laverde, S., Moncada, L. H., Restrepo, A., and Vera, C. L., Mycotic keratitis, 5 cases caused by unusual fungi, *Sabouraudia*, 11, 119, 1973.

447. Zapater, R. C., Albesi, E. J., and Garcia, G. H., Mycotic keratitis by *Drechslera spinifera, Sabouraudia*, 13, 295, 1975.

448. Srivastava, O. P., Lal, B., Agrawal, P. K., Agarwal, S. C., Chandra, B., and Mathur, I. S., Mycotic keratitis due to *Rhizoctonia* sp., *Sabouraudia*, 15, 125, 1977.

449. François, J. and Rijsselaere, M., Corneal infections by Rhodotorula, 1979, *Ophthalmologica*, 178, 241, 1979.

450. François, J. and Rijsselaere, M., Corticosteroids and ocular mycoses: experimental study, *Ann. Ophthalmol.*, 6, 207, 1974.

451. Ley, A. P., Experimental fungus infections of the cornea. A preliminary report, *Am. J. Ophthalmol.*, 42, 59, 1956.

452. Foster, R. K. and Rebell, G., Animal model of *Fusarium solani* keratitis, *Am. J. Ophthalmol.*, 79, 510, 1975.

453. Rysselaere, M., The effect of econazole in experimental oculomycosis in rabbits, *Mykosen*, 24, 238, 1980.

454. Ishibashi, Y. and Matsumoto, Y., Oral ketoconazole therapy for experimental *Candida albicans* keratitis in rabbits, *Sabouraudia*, 22, 323, 1984.

455. O'Day, D. M., Robinson, R., and Head, W. S., Efficacy of antifungal agents in the cornea. I. A comparative study, *Invest. Ophthalmol. Vis. Sci.,* 24, 1098, 1983.
456. Ohno, S., Okumoto, M., Dy-Liacco, J., and Smolin, G., The effect of miconazole on experimental Candida keratitis, *Japn. J. Ophthalmol.,* 20, 438, 1976.
457. Stern, G. A., Okumoto, M., and Smolin, M. A., Combined amphotericin B and rifampin treatment of experimental *Candida albicans* keratitis, *Arch. Ophthalmol.,* 97, 721, 1979.
458. Ohno, S., Fuerst, D. J., Okumoto, M., Grabner, G., and Smolin, G., The effect of K-582, a new antifungal agent, on experimental Candida keratitis, *Invest. Ophthalmol. Vis. Sci.,* 24, 1626, 1983.
459. Wilkie, J., Smolin, G., and Okumoto, M., The effect of rifampicin in Pseudomonas keratitis, *Can. J. Ophthalmol.,* 7, 309, 1972.
460. Smolin, G. and Okumoto, M., Potentiation of Candida albicans keratitis by antilymphocyte serum and corticosteroids, *Am. J. Ophthalmol.,* 68, 675, 1967.
461. Segal, E., Romano, A., Eylan, E., and Stein, R., Experimental and clinical studies of 5-fluorocytosine activity in Candida ocular infections. II. The effect of 5-fluorocytosine in treating experimental Candida infection in rabbits' eyes, *Infection,* 3, 165, 1975.
462. Segal, E., Romano, A., and Barishak, Y. R., Miconazole activity in experimental Aspergillus ocular infections, *Ophthalmic Res.,* 13, 12, 1981.
463. Oji, E. O., Development of quantitative methods of measuring antifungal drug effects in the rabbit cornea, *Br. J. Ophthalmol.,* 65, 89, 1981.
464. Oji, E. O., Ketoconazole: a new imidazole antifungal agent has both prophylactic potential and therapeutic efficacy in keratomycosis of rabbits, *Int. Ophthalmol.,* 5, 163, 1982.
465. Oji, E. O., Mycolase II; an enzyme antifungal agent interacts with the polyene antibiotic pimaricin
466. Oji, E. O. and Clayton, Y. M., The role of econazole in the management of oculomycosis, *Int. Ophthalmol.,* 4, 137, 1981.
467. Jones, B. R. and Al-Hussaini, K., Therapeutic considerations in ocular vaccine, *Trans. Ophthalmol. Soc. U.K.,* 83, 613, 1963.
468. Falcon, M. G. and Jones, B. R., Herpes simplex keratitis: animal models to guide the selection and optimal delivery of antiviral chemotherapy, *J. Antimicrob. Chemother.,* 3(Suppl.), A83, 1977.
469. Shiota, H., Antiviral chemotherapy against herpetic keratitis, in *Proc. XXIII Concilium Ophthalmologicum Kyoto,* Shimizu, K. and Oosterhuis, J., Eds., Excerpta Medica, Amsterdam, 1978, 1754.
470. Komadina, T., Wilkes, T. D. I., Shock, J. P., Ulmer, W. C., Jackson, J., and Bradsher, R. W., Treatment of *Aspergillus fumigatus* keratitis in rabbits with oral and topical ketoconazole, *Am. J. Ophthalmol.,* 99, 476, 1985.
471. Garcia de Lomas, J., Fons, M. A., Nogueira, J. M., Rustom, F., Borras, R., and Buesa, F. J., Chemotherapy of *Aspergillus fumigatus* dermatitis: an experimental study, *Mycopathologia,* 89, 135, 1985.
472. Wahab, S., Lal, B., Jacob, Z., Pandey, V. C., and Srivastava, O. P., Studies on a strain of *Fusarium solani* (Mart.) Sacc. isolated from a case of mycotic keratitis, *Mycopathologia,* 68, 31, 1979.
473. Ellison, A. C., Newmark, E., and Kaufman, H. E., Chemotherapy of experimental keratomycosis, *Am. J. Ophthalmol.,* 68, 812, 1969.
474. Newmark, E., Ellison, A. C., and Kaufman, H. E., Combined pimaricin and dexamethasone therapy of keratomycosis, *Am. J. Ophthalmol.,* 71, 718, 1971.
475. O'Day, D. M., Ray, W. A., Head, W. S., and Robinson, R. D., Influence of the corneal epithelium on the efficacy of topical antifungal agents, *Invest. Ophthalmol. Vis. Sci.,* 25, 855, 1984.
476. Singer, T. A. and Lawton-Smith, T., Experimental corneal histoplasmosis, *Br. J. Ophthalmol.,* 48, 293, 1964.
477. Singh, G., Malik, S. R. K., and Bhatnagar, P. K., *Arch. Ophthalmol.,* 92, 48, 1974.
478. Mahan, M., Sangawe, J. L., and Mahajan, V. M., Pathogenesis of experimentally produced corneal ulcers in rabbits, *Ann. Ophthalmol.,* 16, 246, 1984.
479. Ishibashi, Y. and Matsumoto, Y., Intravenous miconazole therapy for experimental keratomycosis in rabbits, *Sabouraudia,* 23, 55, 1985.
480. Ishibashi, Y., Experimental fungal keratitis due to Fusarium: studies on animal model and inoculation technique. *Proc. XXIII Concilium Ophthalmologicum Kyoto,* Shimizu, K. and Oosterhuis, J., Eds., Excerpta Medica, Amsterdam, 1978, 1705.
481. Ishibashi, Y., Experimental fungal keratitis due to Fusarium. The effects of administration method and dosage of corticosteroid on development of fungal keratitis, *Folia Ophthalmol.,* 29, 149, 1978.
482. François, J., Elenaut-Rysselaere, M., and De Vos, E., Les mycoses oculaires, Masson, Paris, 1968, 33.
483. Graf, K., Über der Einfluss von Kortison auf der Entstehen von Keratomykosen am Kaninchenauge durch saprophytare Pilze des menschlichen Bindehautsackes, *Klin. Mbl. Augenheilk.,* 143, 356, 1968.
484. O'Day, D. M., Ray, W. A., Robinson, R., and Head, W. S., Efficacy of antifungal agents in the cornea. II. Influence of corticosteroids, *Invest. Ophthalmol. Vis. Sci.,* 25, 331, 1984.

485. Oggel, K. and de Decker, W., Standard-Candida-Mykose am Kaninchenauge, *Albrecht v. Graefes Arch. Klin. Exp. Ophthal.*, 193, 193, 1975.

486. Smolin, G., Okumoto, M., and Wilson, F. M., The effect of tobramycin on gentamicin-resistant strains in *Pseudomonas* keratitis, *Am. J. Ophthalmol.*, 77, 583, 1974.

487. Ellison, A. C. and Newmark, E., Effects of subconjunctival pimaricin in experimental keratomycosis, *Am. J. Ophthalmol.*, 75, 790, 1973.

488. Burda, C. D. and Fisher, E., Jr., The use of cortisone in establishing experimental fungal keratitis in rats, *Am. J. Ophthalmol.*, 48, 330, 1959.

489. Ajello, L., Medically important infectious fungi, *Contrib. Microbiol. Immunol.*, 3, 7, 1977.

490. Ajello, L., Dean, D. F., and Irvin, R. S., The zygomycete *Saksenaea vasiformis* as a pathogen of humans with a critical review of the etiology of zygomycosis, *Mycologia*, 68, 52, 1976.

491. Chandler, F. W., Kaplan, W., and Ajello, L., *A Colour Atlas and Textbook of Histopathology of Mycotic Diseases*, Wolfe Medical Publ., London, 1980, 122.

492. Baker, R. D., Leukopenia and therapy in leukemia as factors predisposing to fatal mycoses; mucormycosis, aspergillosis and cryptococcosis, *Am. J. Clin. Pathol.*, 37, 358, 1962.

493. Rippon, J. W., Mucormycosis, in *Medical Mycology: The Pathogenic Fungi and the Pathogenic Actinomycetes*, W. B. Saunders, Philadelphia, 1974, 430.

494. Ainsworth, G. C. and Austwick, P. K. C., Phycomycosis, in *Fungal Diseases of Animals*, 2nd ed., Review Series No. 6 of the Commonwealth Bureau of Animal Health; Commonwealth Agricultural Bureau, Farnham Royal, Slough, 1959, 53.

495. Emmons, C. W., Binford, C. H., and Utz, J. P., *Medical Mycology*, 2nd ed., Lea & Febiger, Philadelphia, 1970.

496. Baker, R. D., The pathologic anatomy of mycoses, in *Handbuch der speziellen pathologischen Anatomie und Histologie*, Ueblinger, E., Ed., Vol. 5 (Part 3), Springer-Verlag, Berlin, 1971.

497. Schofield, R. A., Baker, R. D., and Durham, N. C., Experimental mucormycosis (*Rhizopus* infection) in mice, *Arch. Pathol.*, 61, 407, 1956.

498. Eades, S. M. and Corbel, M. J., Metastatic subcutaneous zygomycosis following intravenous and intracerebral inoculation of *Absidia corymbifera* spores, *Sabouraudia*, 13, 200, 1975.

499. Corbel, M. J. and Eades, S. M., Factors determining the susceptibility of mice to experimental phycomycosis, *J. Med. Microbiol.*, 8, 551, 1975.

500. Eades, S. M. and Corbel, M. J., Enhancement of susceptibility to experimental phycomycosis by agents producing reticuloendothelial stimulation, *Br. Vet. J.*, 131, 622, 1975.

501. Corbel, M. J. and Eades, S. M., Experimental phycomycosis in mice; examination of the role of acquired immunity in resistance to *Absidia ramosa*, *J. Hyg. Camb.*, 77, 221, 1976.

502. Corbel, M. J. and Eades, S. M., The relative susceptibility of NZB and CBA mice to infection with opportunistic fungal pathogens, *Sabouraudia*, 14, 17, 1976.

503. Corbel, M. J. and Eades, S. M., Cerebral mucormycosis following experimental inoculation with *Mortierella wolfii*, *Mycopathologia*, 60, 129, 1977.

504. Corbel, M. J. and Eades, S. M., Experimental mucormycosis in congenitally athymic (nude) mice, *Mycopathologia*, 62, 117, 1977.

505. Smith, J. M. B. and Jones, R. H., Localisation and fate of *Absidia ramosa* spores after intravenous inoculation of mice, *J. Comp. Pathol.*, 83, 49, 1973.

506. Kitz, J. D., Embree, R. W., and Cazin, J., Comparative virulence of *Absidia corymbifera* strains in mice, *Infect. Immun.*, 33, 395, 1981.

507. Waldorf, A. R., Halde, C., and Vedros, N. A., Murine model of pulmonary mucormycosis in cortisone-treated mice, *Sabouraudia*, 20, 217, 1982.

508. Waldorf, A. R., Peter, L., and Polak, A.-M., Mucormycotic infection in mice following prolonged *in vivo* incubation of spores and the role of spore agglutinating antibodies on spore germination, *Sabouraudia*, 22, 101, 1984.

509. Waldorf, A. R., Ruderman, N., and Diamond, R. D., Specific susceptibility to mucormycosis in murine diabetes and bronchoalveolar macrophage defence against *Rhizopus*, *J. Clin. Invest.*, 74, 150, 1984.

510. Waldorf, A. R. and Diamond, R. D., Cerebral mucormycosis in diabetic mice after intrasinus challenge, *Infect. Immun.*, 44, 194, 1984.

511. Symeonidis, A. and Emmons, C. W., Granulomatous growth induced in mice by *Absidia corymbifera*, *Arch. Pathol.*, 60, 251, 1955.

512. Bauer, H., Flanagan, J. F., and Sheldon, W. H., Experimental cerebral mucormycosis in rabbits with alloxan diabetes, *Yale J. Biol. Med.*, 28, 29, 1955.

513. Bauer, H., Flanagan, J. F., and Sheldon, W. H., The effects of metabolic alterations on experimental *Rhizopus oryzae* (mucormycosis) infection, *Yale J. Biol. Med.*, 29, 23, 1956.

514. Reinhardt, D. J., Kaplan, W., and Ajello, L., Experimental cerebral zygomycosis in alloxan-diabetic rabbits. I. Relationship of temperature tolerance of selected zygomycetes to pathogenicity, *Infect. Immun.*, 2, 404, 1970.

515. Kaplan, W., Goss, L. J., Ajello, L., and Ivens, M. S., Pulmonary mucormycosis in a harp seal caused by *Mucor pusillus, Mycopathol. Mycol. Appl.,* 12, 101, 1960.

516. Reinhardt, D. J., Licata, I., Kaplan, W., Ajello, L., Chandler, F. W., and Ellis, J. J., Experimental cerebral zygomycosis in alloxan-diabetic rabbits: variation in virulence among zygomycetes, *Sabouraudia,* 19, 245, 1981.

517. Bauer, H. and Sheldon, W. H., Leukopenia with granulocytopenia in experimental mucormycosis (*Rhizopus oryzae* infection), *J. Exp. Med.,* 106, 501, 1957.

518. Elder, T. D. and Baker, R. D., Pulmonary mucormycosis in rabbits with alloxan diabetes, *Arch. Pathol.,* 61, 159, 1956.

519. Sheldon, W. H. and Bauer, H., The development of the acute inflammatory response to experimental cutaneous mucormycosis in normal and diabetic rabbits, *J. Exp. Med.,* 110, 845, 1959.

520. Sheldon, W. H. and Bauer, H., Activation of quiescent mucormycotic granulomata in rabbits by induction of acute alloxan diabetes, *J. Exp. Med.,* 108, 171, 1958.

521. Mahajan, V. M., Amar, D., and Dayal, Y., Experimental orbital phycomycosis in rabbits, *Mykosen,* 24, 47, 1980.

522. Baker, R. D. and Linares, G., Prednisolone-induced mucormycosis in Rhesus monkeys, *Sabouraudia,* 12, 75, 1974.

523. Martin, J. E., Kroe, D. J., Bostrom, R. E., Johnson, D. J., and Whitney, R. A., Rhino-orbital phycomycosis in a rhesus monkey (*Macaca mulatta*), *J. Am. Vet. Med. Assoc.,* 155, 1253, 1969.

524. Sheldon, W. H. and Bauer, H., Tissue mast cells and acute inflammation in experimental cutaneous mucormycosis of normal, 48/80-treated, and diabetic rats, *J. Exp. Med.,* 112, 1069, 1960.

525. Smith, J. M. B., *In vivo* development of spores of *Absidia ramosa, Sabouraudia,* 14, 11, 1976.

526. Corbel, M. J. and Eades, S. M., Observations on the localization of *Absidia corymbifera in vivo, Sabouraudia,* 16, 125, 1978.

527. Kitz, D. J., Embree, R. W., and Casin, J., Jr., *Radiomyces* a genus in the mucorales pathogenic for mice, *Sabouraudia,* 18, 115, 1980.

528. Howie, J. B. and Helyer, B. J., The immunology and immunopathology of NZB mice, *Adv. Immunol.,* 9, 215, 1968.

529. Cantor, H., Asofsky, R., and Talal, N., Synergy among lymphoid cells mediating the graft-versus-host response. I. Synergy in graft-versus-host reactions produced by cells from NZB/BL mice, *J. Exp. Med.,* 131, 223, 1970.

530. Hart, P. D., Russel, E., Jr., and Remington, J. S., The compromised host and infection. II. Deep fungal infection, *J. Infect. Dis.,* 120, 169, 1969.

531. Marchevsky, A. M., Bottone, E. J., Geller, S. A., and Giger, D. K., The changing spectrum of disease, etiology, and diagnosis of mucormycosis, *Hum. Pathol.,* 11, 457, 1980.

532. Levine, H. B., Cobb, J. M., and Smith, C. E., Immunity to coccidioidomycosis induced in mice by purified spherule, arthrospore and mycelial vaccines, *Trans. N.Y. Acad. Sci.,* 22, 436, 1960.

INDEX

Printed and bound by CPI Group (UK) Ltd, Croydon, CR0 4YY

22/10/2024

01777632-0017